POWER SUPPLIES

POWER SUPPLIES

Jeffrey D. Shepard

Reston Publishing Company, Inc.
A Prentice-Hall Company
Reston, Virginia

Library of Congress Cataloging in Publication Data

Shepard, Jeffrey D.
  Power supplies.

  Bibliography: p.
  Includes index.
  1. Electronic apparatus and appliances--Power supply.
I. Title.
TK 7868.P6S48 1984   621.319'24   83-21190
ISBN 0-8359-5568-0

© 1984 by **Reston Publishing Company, Inc.**
  A *Prentice-Hall Company*
  Reston, Virginia 22090

10   9   8   7   6   5   4   3   2   1

Interior design and production: Jack Zibulsky

Printed in the United States of America

**To Peg with love.
Thank you for your patience.**

CONTENTS

PREFACE

This book grew out of my experiences in the
marketing department of a major manufacturer
of switching regulated power supplies. I have
seen the specific questions and problems electronic engineers have when the time comes to
determine how to deliver the power needed by the circuits of a newly designed
system. This book provides practical technical information you need for the
selection, specification, and application of power supplies.

The dynamic nature of today's electronic technology demands that you
spend whatever time you can spare simply keeping abreast of developments in
your own field. There is rarely any spare time available to pursue secondary areas
such as power supplies. As a result, most engineers have to rely on past experience, not current technical information when working with power supplies.

On the surface power supply technology is not a rapidly changing area.
After all, you still order power supplies based on volts and amps. However, since
power supply technology is changing in subtle ways, effective power supply
specification is no longer a matter of ordering '5 Volts at 30 Amps'. It has

become a much more complex process with selections to be made between linears and switchers, flybacks and half-bridges, and semi-regulated and fully regulated equipment.

While not providing a detailed theoretical discussion of power supply design considerations, this book provides a sound theoretical basis from which you can proceed with the selection, specification, and application process. Good engineering cannot be reduced to a simple set of rules. I have not attempted to set up rules for power supply useage, but have provided the information you will need to develop your own selection criteria. Once you have the technical background to become independent of specific rules, you will be free to seek creative solutions to your power supply requirements.

INTRODUCTION

1

Most engineers are involved in power supply selection, specification, and application—not power supply design. While there are many sources of power supply design information, from textbooks to trade publications to application notes provided by component manufacturers, until now there has not been a comprehensive source of practical information on power supply selection, specification, and application. This book is not intended to be a heavy design treatise, but rather a practical source document for the practicing engineer.

Proper selection is the first step in reliable power supply and system operation. To select properly, the user must know as much as possible about the operating characteristics of the various regulation techniques. Furthermore, he or she should know the advantages and disadvantages of specific circuit configurations, their behavior under various operating conditions, their construction, their effect on circuits, and the effect of circuits on them. Chapters 2 and 3 discuss the function and implementation of the three primary regulation techniques: ferroresonant, linear, and switching in terms of selection criteria.

1

Once the general regulation approach has been identified, it is necessary to develop a detailed specification of the precise power supply required by a particular application. Chapter 4 focuses on the preparation of complete power supply specifications. Structured as a paragraph by paragraph review and comparison of all key specifications for ferroresonant, linear, and switching power supplies, this chapter is the specifying engineer's "do-it-yourself" guide to the preparation of highly detailed power supply specifications.

Insuring that the power supply delivered by the vendor meets the operating parameters defined in the specification in a reliable manner is the next major consideration. Power supply reliability and testing are the topics of Chapters 5 and 6. Beginning with a discussion of mathematical reliability predictions and MIL-HDBK-217, the focus quickly shifts to a practical analysis of thermal design considerations and ends with a "how-to" presentation of testing and evaluation procedures for all key power supply operating parameters. Also discussed is the growing use of automatic test equipment in power supply manufacturing and its effects on power supply specifiers and users.

Continuing the theme of practical power supply applications information, Chapter 7 covers protection circuits, power distribution techniques, and power supply paralleling for increased power or redundant operation. Providing power and protection to the working circuits of electronic systems involves making a number of interrelated choices regarding power supply and system design considerations. The interrelationships involved are discussed on a practical level as they relate to system performance.

Agencies outside the mainstream of the electronics industry are having more and more to say about the design of electronic systems. The issues involved in complying with the growing number of government and industry safety and electromagnetic interference standards (UL, CSA, FCC, VDE, IEC, etc.) are presented in Chapter 8. Compliance with the European standards is a new experience for many engineers. These newly discovered system design requirements are presented, and their implications for power supply specifiers are discussed.

Switching regulated power supplies are becoming more and more common in all types of electronic systems. Chapters 9 and 10 discuss some of the "side-effects" of the increasing use of switching power supplies. A technical and financial analysis of the benefits and pitfalls of making versus buying power supplies, with a heavy concentration on switching types, is the subject of Chapter 9. New developments in power supply technology, specifically the rapid development of switching regulation techniques, are reviewed in Chapter 10. The discussion focuses on the dynamic nature of power supply technology in terms of price–performance changes resulting from new components and design techniques, and makes some predictions about future developments and their implications for power supply specifiers and users.

Finally, there is a comprehensive glossary of 148 power supply terms. The glossary includes references to the body of the book as well as definitions for each of the terms listed. When used in conjunction with the index, the glossary makes the practical information contained in this book easily accessed.

All electronic systems require a power supply. Proper selection, specification, and application of power supplies is an important aspect of the engineer's function. Unfortunately, many engineers have not had access to all the information necessary to satisfy their power supply needs in an efficient manner. The increasing complexity of electronic systems in general, and power supplies in particular, is placing more demands than ever on the always scarce resources of engineering. This book was written to provide engineers with a complete source of practical power supply information from the basics of their operation to comprehensive applications information.

POWER CONVERSION TECHNIQUES

The majority of electric power used to supply today's sophisticated electronic systems comes from what are termed *primary* power sources. These primary sources include hydroelectric, coal- and oil-fired, and nuclear electric power generators. Their prime objective is to produce large quantities of electric power in the most economic fashion possible. The quality of the electricity is a secondary consideration.

Medium-quality electric power is fine for operating common lights or motors; however, it leaves much to be desired when it comes to powering electronic systems. This is where power conversion comes into play. In electronic systems, power conversion is necessary to screen out voltage fluctuations, electric noise, and other undesirable parameters, as well as to change the voltage and/or frequency to other values as required by various electronic systems.

EQUIPMENT CLASSIFICATION

Power conversion equipment can be divided into four broad classifications, as discussed in the following.

AC in/AC out

There are two types of ac to ac power conversion equipment. *Frequency changers* do just what their name states. They transform the electric power to a different frequency. For example, some mainframe computers operate on 415 hertz (Hz) ac and employ a frequency changer to change the 60-Hz nominal frequency of the ac line to 415-Hz power for the system.

Line regulators operate at the same frequency, but change the degree of regulation, filter noise, and/or the voltage of the incoming ac power. They are being used more and more to shelter electronic systems from what is increasingly "dirty" ac input power.

DC in/AC out

Inverters are employed to provide ac power when the primary power source is dc. They are commonly used in all types of mobile applications when the primary power source is a bank of batteries. Applications include inverters that provide the 400-Hz power found on aircraft and aboard ships and those that provide 60-Hz power from a car or truck battery for the remote operation of ac-powered equipment.

DC in/DC out

A dc to dc *converter* is used to change the voltage and improve the regulation of bulk sources of dc electric power such as batteries. Typical applications include remote telecommunication and mobile systems where 48, 60, or even 120 V dc is provided by the storage batteries and must be converted before it can be used by the electronic system.

AC in/DC out

Power supply is the common term for electronic devices that provide dc output voltages from an ac primary source. All of today's solid-state electronic systems require dc operating voltages, and in most cases these dc voltages are provided by a power supply. For this reason, it is important that every electronics engineer have a basic understanding of power supply principles.

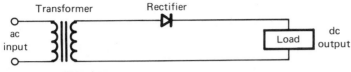

FIGURE 2.1.
A brute-force power supply employs two
functional elements to convert the ac line
voltage to an unregulated dc output voltage.

POWER SUPPLY
OPERATION

The most basic form of general-purpose power supply is the *brute-force* supply. As
the name implies, this type of power supply is very simple and provides an
unregulated dc output voltage. A brute-force supply consists of two functional
elements (Figure 2.1), the *power transformer* and the *rectifier*.

The ac line voltage is connected across the primary side of the power
transformer. The secondary windings produce an ac voltage of approximately
the value required by the devices to be powered. This secondary ac voltage is
converted to dc by the rectifier. The *load* (system to be powered) is connected
across the dc output provided by the rectifier.

The rectifier sections of all power supplies, brute force or regulated, convert the transformer's ac output into pulsating dc voltage. This is accomplished
using diodes that conduct current in one direction and block current flow in the
other. There are three basic rectification techniques: half-wave, full wave, and
bridge.

The simplest rectification design is the *half-wave form*. Half-wave rectification is accomplished by placing a single diode in series with the load. Current in
the transformer's coils reverses direction during each half-cycle of ac. The diode

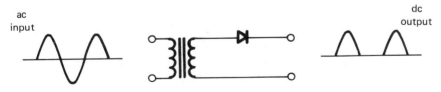

FIGURE 2.2.
Half-wave rectification produces dead time
between the current pulses, causing filtering difficulties.

can only conduct current in one direction and therefore blocks one-half the ac wave-producing pulses of current with dead time between the pulses (Figure 2.2). The dead time, which is a part of half-wave rectification, causes filtering difficulties and uses only half of the ac wave.

Full-wave rectification, as its name states, uses the entire ac wave, thereby reducing the filtering problems. To accomplish this, a more complex center-tapped transformer and an additional diode are generally used (Figure 2.3). One diode conducts for one-half of the ac cycle, and the other conducts on the second half. The two diodes alternately rectify the ac wave, eliminating the dead time between the pulses of rectified current. The fact that center-tapped transformers are larger and more expensive is offset by this design's superior efficiency and reliability.

The need for a center-tapped transformer can be eliminated by using four diodes as in the bridge rectifier (Figure 2.4). A *bridge rectifier* uses the same transformer as a half-wave rectifier but eliminates the dead time between the current pulses, just like the full-wave type. A primary consideration in choosing between the full-wave, center-tapped and bridge designs is the trade-off in magnetics and semiconductor costs. At low power levels, the two additional diodes are less costly than, and just as reliable as, a center-tapped transformer. Therefore, bridge rectification is often used in low power applications. As the power level increases, the relative cost of diodes goes up and their reliability goes down compared to a center-tapped transformer; thus at higher power levels, full-wave rectification is used.

Brute-force power supplies have a limited application in electronic systems. Any changes in line voltage are proportionally transmitted through the power transformer to the output. While this lack of line regulation can be a problem, the brute-force supply's lack of load regulation is even more of a problem in most systems. While the ac input line may vary as much as 20% in some cases, it is not unusual for the load to vary 25%, 50%, or even 100%.

Lack of dynamic load regulation is the most important weakness of the brute-force approach. It is a result of the relatively high dynamic impedance of brute-force supplies. The transformer windings and rectification diodes both

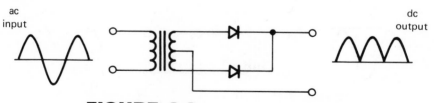

FIGURE 2.3.
Full-wave rectification uses two diodes and a center-topped transformer to eliminate the dead time between current pulses.

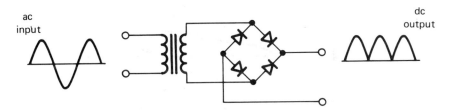

FIGURE 2.4.
Bridge rectification uses four diodes to eliminate the dead time between current pulses.

exhibit significant impedances. As the level of load current rises, the voltage
drops across these components rise, and the effective output voltage drops. The
reverse occurs during drops in the load current, causing the output voltage to
rise.

A brute-force power supply provides a low-quality (unregulated) dc output
voltage. Changes in line voltage and/or load current will result in changes in the
output voltage. In addition, even under constant line and load conditions, the
output of this type of power supply is not pure dc but contains a large amount of
ripple (ac-related fluctuations). Most sophisticated electronic systems cannot
operate on unregulated dc. When a constant output voltage is required, this
brute force supply must be regulated and filtered.

REGULATION

A regulated power supply maintains a steady, constant output voltage by sensing
its output and automatically compensating to eliminate any changes. There are
three basic regulation techniques, each with a unique set of strengths and
weaknesses.

Ferroresonance

Ferroresonance is the simplest approach to voltage regulation. This type of
power supply relies on a *constant-voltage* transformer in which the secondary of
the transformer is operating in saturation. By operating in saturation, the flux
density in the secondary and hence the output voltage remains relatively constant, regardless of the input voltage (Figure 2.5).

The operation of a ferroresonant power supply is almost completely
dependent upon the design of the transformer. The iron core is divided into two
magnetic circuits separated by a magnetic shunt (Figure 2.6) so that the secondary magnetic circuit can be put into saturation without having the primary go
into saturation. The method by which this is accomplished is to incorporate a

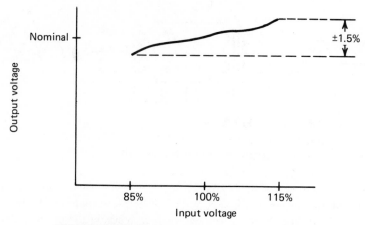

FIGURE 2.5.

Ferroresonant regulation can deliver line regulation due to the inherent isolation between the primary and secondary windings.

resonant tank circuit in the secondary. The tank circuit is "tuned" to the source frequency (typically 60 Hz) so that a resonance is set up between C1 and the secondary of the transformer. The resonance involved is not an inductive–capacitive resonance since the inductor in this case is saturated. It is the resonance of the core's ferromagnetization characteristic with capacitor C1. This is the source of the term *ferroresonance*.

The output of a ferroresonant transformer has a voltage wave form that is semisquare and is well suited to rectification and filtering (Figure 2.7). Com-

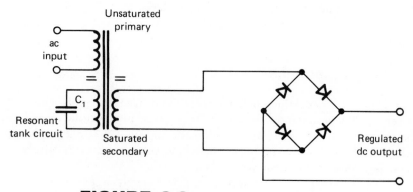

FIGURE 2.6.

Ferroresonant voltage regulation is derived from a specially designed constant-voltage transformer.

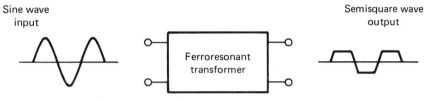

Sine wave
input

Ferroresonant
transformer

Semisquare wave
output

FIGURE 2.7.
The semisquare-wave output voltage of a
ferroresonant transformer is well suited to
rectification and filtering.

pared to a sine wave, this type of semisquare wave gives significantly better load
stabilization and lower ripple when filtered. Better load stabilization results from
the lower peak-to-average ratio, while the lower ripple is a consequence of the
shorter discharge time period compared to a sine wave.

Inherent current limiting is an important feature of a ferroresonant power
supply. When overloaded, the Q (*quality factor*) of the resonant tank circuit falls
below the level necessary to continue oscillation, and the output drops. Excessive current demand greatly diminishes the mutual flux between the primary and
secondary windings. This decreases the effective energy coupling (the Q) between the input and output, causing the secondary voltage to rapidly fall to zero
(Figure 2.8). Another important aspect of this type of power supply is the
inherent absence of overvoltage risk, without the use of other voltage-limiting
techniques.

Because ferroresonant supplies are nondissipative in operation (with the
exception of minimal core losses and the like), they are very efficient. Typical

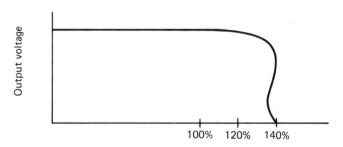

Output voltage

100%   120%   140%

Load current

FIGURE 2.8.
Inherent current limiting in ferroresonant
power supplies results from the collapse of
output voltage as load current becomes
excessive.

efficiency will run around 80%. Their low component count, together with no need for overcurrent or overvoltage protective circuitry, means that ferroresonant power supplies are exceptionally reliable.

The ferroresonant approach to voltage regulation is not without drawbacks. Ferroresonant transformers and hence power supplies tend to be rather bulky and very heavy. Because of the long time constant of the resonant circuit, they also tend to have unusually long transient response times. Although they provide very good line regulation, their load regulation tends to be less accurate than other approaches.

Since the successful operation of a ferroresonant transformer depends on having the source voltage frequency very near the resonant frequency of the tank circuit, power supplies based on this technique are limited in terms of input frequency range. A typical ferroresonant supply will be specified for operation over a narrow frequency range such as 60 Hz $\pm$ 0.5%, since a 1% change in frequency will result in approximately a 1.5% change in voltage.

On the other hand, ferroresonant supplies are not overly sensitive to changes in input voltage. Since the secondary is operated in saturation, a large change in the input voltage is required before any change is passed through to the load. These power supplies can easily tolerate input voltages of $\pm$15%.

Linear power supplies

Linear power supplies (*linears*) do not depend upon specially designed magnetics for voltage regulation. They are comprised of a basic brute-force power supply, an electronic variable resistance element (the pass transistor), a voltage detector, and a stable reference voltage (Figure 2.9). In ferroresonant supplies, the regulation occurs in the ferroresonant transformer prior to rectification. In a linear, regulation occurs after rectification.

The voltage detector constantly monitors the dc output voltage and compares it to the reference voltage. If the output voltage begins to change, whether due to a change in line voltage or load current, the detector senses this and immediately adjusts the voltage drop across the pass transistor as necessary to ensure the desired constant dc voltage across the output.

Such linear feedback circuits can operate very quickly, typically from 10 to 50 microseconds (μs). They have a very high gain for a high degree of regulation and can be controlled over a wide range of outputs from cutoff (zero volts out) to near saturation (maximum voltage out). Relatively low efficiency (30% to 45%) is one of their primary shortcomings.

The efficiency of a linear supply increases somewhat at higher output voltages since the head room needed from the secondary of the transformer is reduced as a proportion of the output voltage. To provide a regulated 5-volt (V) output, the transformer may have to provide 6 to 7 V to the pass element. The difference between the 5-V output and the 6 to 7 V supplied by the transformer

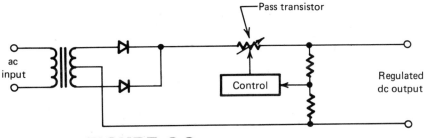

FIGURE 2.9.
Linear regulation is accomplished by controlling the voltage drop across the variable resistance element (the pass transistor).

is commonly called the *head room*. A linear supply producing a 12-V output may need 13 to 14 V from the secondary of the transformer. This proportionally lower head room also means a smaller relative voltage drop across the pass element and somewhat higher efficiency.

The need for the head room comes from the requirement that the supply be able to operate at full load and low line voltage simultaneously. The lower the specified low line voltage, the higher the head room at nominal line. This is the primary reason that linear power supplies have relatively narrow input voltage bands (±10%).

Because of the relative inefficiency of linear power supplies, they generally require extensive cooling. The heat sink for the pass transistors is typically one of the larger components in a linear. Another large component is the relatively massive 60-Hz transformer that is used to interface between the input line and the rectifier section. It is not unusual for the transformer of a medium-power linear to weigh 50 pounds. The net result of the large 60-Hz magnetics and the need for heat dissipation is that linears tend to be fairly large and heavy, although not so much so as ferroresonants.

Switching regulation

Switching regulated power supplies are nondissipative and, like ferroresonants, offer high efficiency (typically 70% to 80%). However, unlike the relatively passive nature of the ferroresonant regulation technique, switchers are like linears in that they employ active, feedback-oriented regulation. In a typical off-line switcher, the regulation occurs on the primary side of the power transformer rather than on the secondary side, as is the case with linears.

In a typical switcher, the control circuit monitors the dc output voltage and compares it to a stable reference voltage. The control circuit determines the ratio of on-time to off-time of the electronic switch connected in series between

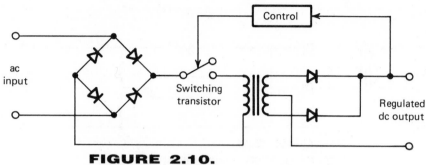

FIGURE 2.10.
Switching regulation is accomplished by controlling the ratio of on-time to off-time of the switching transistor.

the input rectifier and the power transformer (Figure 2.10). Increasing the amount of time the switch remains closed increases the output voltage, whereas decreasing the closure time decreases the voltage. The control circuit varies the relative amount of on-time to off-time to maintain a steady output voltage, regardless of variations in input line voltage or output load current (Figure 2.11). This technique is known as *pulse-width modulation* (PWM) and is most commonly used at frequencies above 20 kilohertz (kHz).

The frequency of operation is selected as a compromise. As the switching frequency is increased from 60-Hz line frequency, magnetic sizes go down and semiconductor switching losses go up. At higher frequencies, even wiring induc-tance becomes a critical factor. Switching supplies are typically operated be-

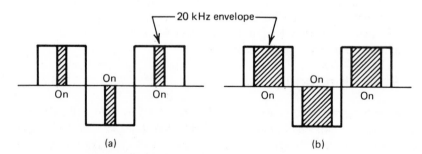

FIGURE 2.11.
In a PWM switcher, the pulse width is (a) narrow for light loads, and (b) wide for heavy loads, but always remains within the envelope of the primary switching fre-quency.

tween 20 and 50 kHz, which is high enough to eliminate any potential audible noise, but still low enough to keep switching losses at reasonable levels.

A second approach to switching regulation operates at line frequency and employs silicon-controlled rectifiers (SCRs) in place of the switching transistors used in the pulse-width approach. Called *phase-controlled modulation,* the control circuit in this type of switching regulator determines the phase angle at which the SCRs are turned on to provide more or less energy to the output as required by the load (Figure 2.12). When operating under light loads, the SCRs are fired (turned on) late in each half-cycle of ac input; under heavy loads, they are fired earlier during each half-cycle.

Phase control is less costly, but involves a number of specification trade-offs, including less energy storage capacity than PWM switchers, higher noise and ripple (up to 500 millivolts (mV) peak to peak) due to the variable phase components of the output, and poorer transient response and regulation, since, like ferroresonants, the phase-control switcher must go through a number of cycles at line frequency before responding to changes in input voltage or output current.

Advantages of the phase-control approach include high efficiency and excellent reliability, in addition to low cost. In most applications, however, the disadvantages of phase-controlled switching outweigh the benefits and, as a result, it has a very limited application in electronic systems. For the balance of this book, the terms *switching regulation* and *switcher* will refer exclusively to the pulse-width-modulation type of power supply.

Because switchers operate at high frequencies, they employ much smaller transformers and filter components than are found in linear or ferroresonant designs. In addition, switchers are highly efficient (70% to 80%) and require substantially less heat sinking than a linear supply of equal output. As a result of its much smaller components, a switcher will typically be about one-third to one-sixth the size of an equivalent linear or ferroresonant supply. The high

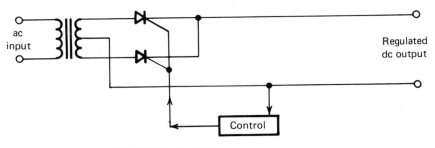

FIGURE 2.12.
Phase-control modulation uses silicon-controlled rectifiers to perform both regulation and rectification.

efficiency of switchers results from the fact that, when the transistors are on, they present very little resistance. The power is primarily dissipated in the load and, unlike linears, not in the regulation process.

High-frequency switching is the source of most of the supply's problems, as well as its attributes. By operating at or above 20 kHz, switchers can generate a significant amount of electromagnetic interference (EMI), and special attention must be given to filtering this unwanted characteristic. In addition, high-frequency, high-power switching places heavy stresses on the switching transistors, and good, conservative design practices in the switching section are especially important to ensure reliable operation.

The switching section of a typical switcher operates at 300 V dc. It is this high-voltage operation that gives switchers the ability to carry over operation through line voltage dropouts. This carry-over, or holdup, is due to the fact that the energy storage in a capacitor is proportional to the product of the capacitance and the square of the voltage ($E \propto CV^2$). As a result, a switcher (300 V dc) can store many times as much energy as a linear (5 V dc).

FILTERING

Whatever form of regulation is used, it is also generally necessary to include filters in a power supply to smooth out fluctuations that remain after the rectification and regulation processes. The final result is a steady, regulated, "pure" dc voltage at the output of the power supply.

In simple capacitive filtering (Figure 2.13), a capacitor is connected in parallel across the output of the power supply. The pulsing voltage from the rectifier and regulation sections charges the capacitor to a voltage that is equal to the peak voltage of the pulsations. The capacitor then acts as a reservoir, absorbing and discharging energy as necessary to provide a constant dc output voltage to the load.

Regulated
dc

Filtered
dc output

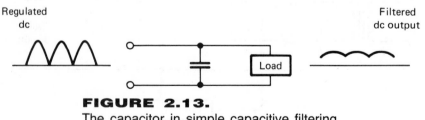

FIGURE 2.13.
The capacitor in simple capacitive filtering acts as an energy reservoir absorbing and discharging energy as necessary to provide a filtered dc output voltage to the load.

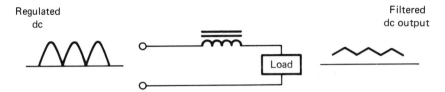

Regulated
dc

Filtered
dc output

Load

FIGURE 2.14.
Simple inductive filtering relies on the electromagnetic field that builds up around the
inductor to act as a reservoir to provide a
relatively constant dc output voltage to the
load.

A second filtering method is to connect an inductor (a magnetic coil with
low dc resistance) in series with the load (Figure 2.14). The current flowing
through the inductor builds up an electromagnetic field that induces voltage
into the inductor. The induced voltage acts as an energy reservoir to provide a
constant dc output voltage to the load, regardless of pulsations of the current
from the rectifier and regulation sections of the power supply.

In most practical applications, the filtering produced by either a single
capacitor or inductor is insufficient, and a more complex filter, known as an *LC*
filter, must be used (Figure 2.15). The capacitor (a current device) and the
inductor (a voltage device) work in cooperation to produce an extremely high
quality, well-filtered output voltage. The complexity of the filter network depends upon the type of regulation used and the sensitivity of the load to fluctuations in the output voltage. Switchers tend to require more extensive filtering
than either ferroresonants or linears, and analog circuits must have "purer" dc
voltages than logic circuits.

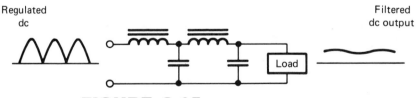

Regulated
dc

Filtered
dc output

Load

FIGURE 2.15.
Complex inductive–capacitive (*LC*) filter
networks are used to provide an extremely
high quality, well-filtered dc output voltage
to the load.

SUMMARY

With the exception of the brute-force type, all power supplies incorporate similar rectification and filtering functions. The most basic differences in regulated power supplies are in the regulation techniques employed. This chapter has focused on the basic technologies available (ferroresonant, linear, and switcher) for converting medium-quality ac power into high-quality, regulated dc power for sophisticated electronic systems.

Significant trade-offs are involved in selecting a regulation technique (Table 2.1). Ferroresonant power supplies tend to be highly efficient and reliable, but they are large, heavy, and very sensitive to changes in input frequency. Linears are not as sensitive to changes in input frequency and provide better regulation, but they are less efficient and are more sensitive to changes in input voltage. Switchers are small, light, efficient, and can provide constant power through dropouts in the ac input voltage, but they are complex, have a slower transient response than linears, and can produce high-frequency noise, which requires additional filtering.

Three general parameters can be used to compare regulation techniques: relative cost per watt, regulation versus transient response time, and power density (watts per cubic inch; W/in.^{3}) (Figure 2.16). Linears yield medium relative cost, excellent regulation, and the lowest power density. Ferroresonants offer the lowest relative cost and medium power density at the sacrifice of regulation, since they have the poorest regulation. Switchers, on the other hand, have best power density and average

TABLE 2.1
Regulation Technique Trade-Offs

Parameter	Ferroresonant	Linear	PWM
Level of technology	Mature	Mature	Developing
Power density (W/in.^{3})	$\frac{1}{6}$	$\frac{1}{3}$	1+
Watts per pound	3	5	20
Efficiency (%)	75	40	75
Input voltage (%)	±15	±8	±20
Input frequency (Hz)	60	50–63	47–440
Holdup time (ms)	1	1	20
Regulation (%)	3	0.05	0.2
Noise (mV)	160	10	50
Primary noise frequency (Hz)	60	60	20K

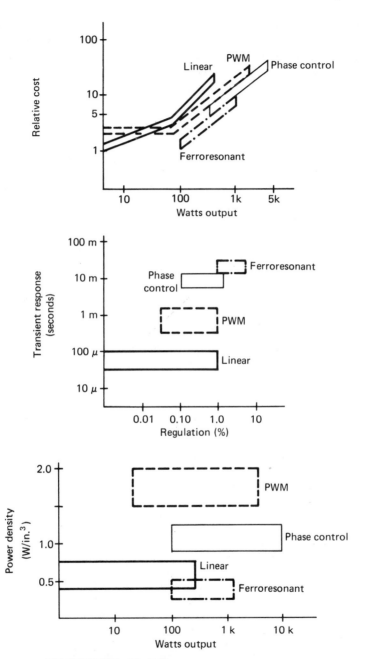

FIGURE 2.16.
Comparison of key regulation parameters.

regulation, but they can cost more at power levels under 100 W. At power levels between 50 and 100 W, linears, ferroresonants, and switchers are similar in cost. Above 100 W, switchers and ferroresonants are generally more cost-effective, with switchers employed in the vast majority of applications owing to their better overall operating characteristics.

CIRCUIT CONFIGURATIONS

Each of the basic regulation techniques, ferroresonant, linear, and switcher, offers inherent advantages and disadvantages. Equally
important with the regulation technique is the
specific circuit configuration. Each technique
may be implemented with a number of different circuit configurations, each
with its own pluses and minuses. Additional variations are possible by combining certain regulation techniques in a single unit.

The situation can become further complicated when multiple-output
power supplies are considered. While it is unusual for a single output unit to
incorporate more than two regulation techniques, a multiple-output unit may
employ three, four, or more forms of regulation. Each output may power a load
that has significantly different requirements. Examples are motors, which can
tolerate semiregulated voltages, digital circuits, which must have tighter regulation, and analog circuits, which demand very tight regulation.

FERRORESONANTS

The simplest form of ferroresonant supply (ferro) is the single primary version. In a ferroresonant or constant-voltage transformer (CVT), the flux in the primary magnetic circuit is carried by unsaturated core material, as in a conventional transformer. The secondary flux, however, is carried by saturated core material. The result is that a change in the primary flux (caused by a change in input voltage) produces far less than a proportional change in secondary flux or output voltage.

The key to the operation of a ferroresonant transformer is the magnetic shunt (Figure 3.1). The shunt decouples much of the secondary flux from the primary winding. The tank circuit capacitor, C1, causes a large reactive current in the secondary winding, causing the saturation of the secondary core material. The saturation increases the isolation between the primary and secondary, since more primary flux travels through the shunt rather than the secondary core. Perfect isolation would leave the secondary separated from its source of energy in the primary circuit and would prevent any energy transfer from input to load. Since there is not complete isolation between primary and secondary, and because the transition from unsaturated primary to saturated secondary is gradual, not sharp, the regulation provided by single primary ferroresonants is limited.

There are two ways to improve the performance of a ferroresonant power supply. It is possible to add a second, *compensating winding* to the primary side of the transformer, which will provide a feedback mechanism and thereby improve line regulation (Figure 3.2). This winding carries the unrectified load current and is wound to oppose the primary flux. The feedback mechanism in this case is direct and involves no active components.

An increase in line voltage would, initially, cause a correspondingly higher secondary voltage, which tends to produce a larger unregulated load current

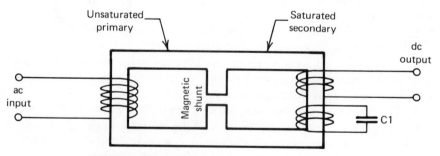

FIGURE 3.1.
The magnetic shunt provides decoupling between the primary and secondary flux circuits and is the key to ferroresonant regulation.

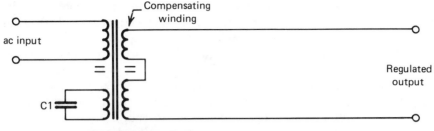

FIGURE 3.2.
A compensating winding added to the primary side of a ferroresonant transformer provides a direct feedback mechanism that improves line regulation.

flowing through the compensating winding, causing decoupling in the primary and, thereby, reducing the voltage in the secondary. Addition of a compensating winding significantly improves line regulation (by at least a factor of 2) without seriously increasing the design complexity or adding components.

Ferroresonant transformers generally provide a 10-to-1 improvement over variations in input voltage (line regulation) and give load regulation on the order of ±2%. Load regulation can be improved by what is termed *postregulation* of the output voltage. The postregulator is typically located between the rectifier section and the output filter (Figure 3.3) and usually consists of a linear regulator. By combining ferroresonant and linear regulators in this way, it is possible to realize some of the best features of both.

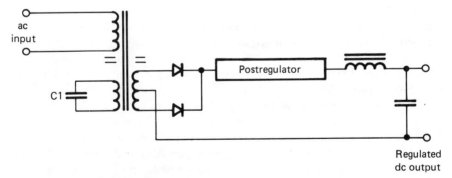

FIGURE 3.3.
Combining a ferroresonant preregulator for line regulation and a linear postregulator for load regulation will yield some of the best features of both.

Stated simply, the ferroresonant handles most line regulation, while the linear takes care of load regulation. Efficiency is much better than with a straight linear since it is now possible to implement a linear postregulator design optimized to a relatively narrow band of input voltages owing to the preregulation provided by the ferroresonant. Much of the inefficiency of linears results from the need to operate under varying line conditions. The ferroresonant eliminates most line variations and allows the linear to be designed for a narrow input voltage band with a resulting improvement in efficiency.

The efficiency of this hybrid regulation approach is between that of either a pure linear or ferroresonant. Other characteristics of this hybrid design are improved transient response time and load regulation from the linear postregulator, and the inherent current-limiting and overvoltage protection provided by the ferroresonant preregulator.

This hybrid approach is not without its drawbacks. It inherits the ferro's sensitivity to input voltage frequency, tends to be somewhat bulky, consisting as it does of both a large constant-voltage transformer and heat sinking for the dissipative series-pass element, and can be more costly than either a simple ferro or linear supply.

LINEARS

Linear power supplies fit into two general classifications: they employ either a *series regulator* or a *shunt regulator* (Figure 3.4). The operating specifications of these techniques are similar, with the following significant exceptions: (1) The electrical efficiency of shunt regulators is directly related to the relative loading on the output. As a result, shunt regulators are generally not employed in systems that operate at less than half their peak load for extended periods of time. While the efficiency of series regulators is also related to relative loading, they are not as sensitive as shunt regulators. (2) The shunt regulator is protected from overload and short circuit without the auxiliary circuitry needed by the series regulator. (3) The shunt regulator provides a path for reverse current and will absorb current from the load as well as deliver current to it. This can be especially useful in situations involving inductive loads.

Series regulation is, by far, the most common form of linear power supply. Two series regulation control techniques are generally employed: *zener reference control* and *feedback control* (Figure 3.5). Series regulators using a zener reference are more efficient, provide much more load current, and better, though still poor, regulation than a shunt regulator. In such a supply, a common-base power transistor dissipates the difference between input and output voltages. In its simplest form, a zener diode absorbs base current, which is much less than load current and regulates the voltage drop across the pass transistor. The zener diode and the *NPN* transistor, used for controlling series transistor base drive, have

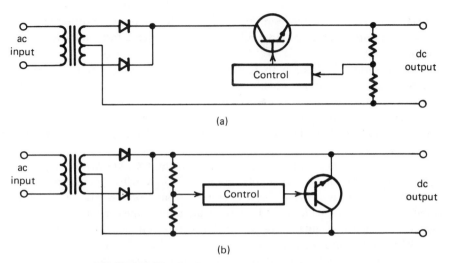

(a)

(b)

FIGURE 3.4.

Both (a) series and (b) shunt type linear regulators use active, dissipative regulating action.

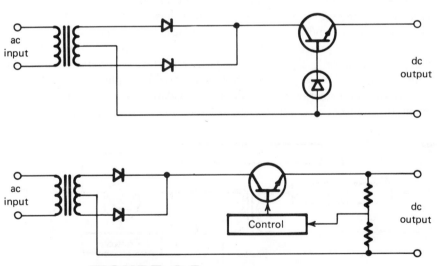

FIGURE 3.5.

(a) Series regulators with zener reference control provide improved regulation over the shunt approach; (b) the feedback control technique provides even more improvement in regulation.

temperature coefficients that tend to balance each other out, resulting in a low net temperature coefficient and relatively good temperature stability of the output voltage.

In a series regulator with feedback control, a voltage-divider network inserted across the output provides a feedback voltage to the control circuit, which is compared with a reference. The output is adjusted (to make them equal) by varying the base drive current of the series power transistor. This high-gain feedback loop greatly improves load regulation over the zener reference technique.

Many "cookbook" linear regulator circuits are well known today. Linear regulation is a mature power supply design technique with specialized circuits developed to improve current-limit operation, transient response, and overall stability. Once only made from discrete components, linear regulators are now available in highly reliable, three-terminal integrated circuits that can deliver up to 5 W of power and regulate output voltage to within 5 mV.

Whichever linear regulation technique is employed, the efficiency of a linear power supply may be improved by adding *preregulation*. This was briefly discussed in Chapter 2, where the preregulation was performed by a ferro and the postregulation was linear. Also called *double regulation*, preregulation can be employed on either the primary or secondary side of the power transformer. In either case, with the exception of ferroresonant preregulation, the preregulator senses the voltage drop across the postregulator and adjusts to ensure that a relatively constant and low voltage drop is maintained (Figure 3.6).

Primary side preregulation can incorporate SCR phase control (a form of switching regulation), a preregulating pass transistor, or a variable-amplitude oscillator. Use of SCR phase control or a variable-amplitude oscillator can improve overall regulation by providing line regulation, but it may also extract a penalty in the form of lower reliability due to increased component count and

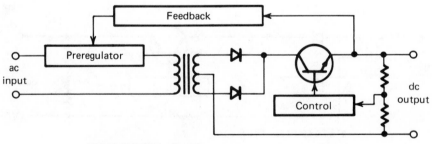

FIGURE 3.6.
Preregulation schemes sense the voltage drop across the pass transistor and adjust to ensure that a relatively constant and low voltage drop is maintained.

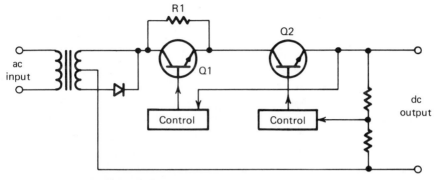

FIGURE 3.7.

In secondary preregulation schemes, over-all efficiency is not greatly improved, but less expensive pass transistors can be used since the total power dissipated in the regulation process is shared by Q1, Q2, and R1.

greater circuit complexity. A preregulating pass transistor avoids most of the reliability penalty, but it does not appreciably improve overall efficiency.

A preregulating transistor is typically found in secondary preregulation schemes (Figure 3.7). The major advantage of this approach is that lower-power pass transistors can be employed for the same load rating, since each dissipates about one-quarter the power that would have to be dissipated by a single-pass element. However, efficiency is not significantly increased since most of the dissipation difference is now seen by resistor R1.

Preregulation schemes always mean added components and increased complexity. Not all preregulation techniques improve the overall efficiency of linears. They do, however, allow smaller and less expensive final regulators to be used. This can result in the benefit of a lower-cost supply if the number of additional components required for preregulation is minimized so that their added cost does not overshadow the cost savings realized from using less expensive final regulators.

SWITCHERS

Switching regulated power supplies are much more complex than either linears or ferros. Out of this complexity comes a power supply with high efficiency, high power density, and good line and load regulation. Regulation in switching supplies generally takes place on the primary side of the main power transformer (Figure 3.8).

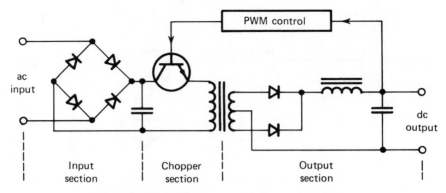

FIGURE 3.8.

Pulse-width-modulation type of switching regulation occurs on the primary side of the power transformer and involves feedback from the secondary side to the primary side.

In off-line switchers (the most common type), the input rectifier and filter section is situated directly across the ac input line. This section performs four primary functions: (1) voltage rectification and doubling, (2) energy storage, (3) inrush current limiting at turn on, and (4) supression of electromagnetic interference.

The ac input is first rectified and, in the case of 115-V units, doubled to 300 V dc. The power conversion occurs in the chopper section at 300 V dc and a correspondingly low current depending on the output load. The 300 V dc is used to charge a bank of large capacitors that actually feed the chopper section. These capacitors provide the energy for the holdup time of the power supply.

The 300-V input capacitors are sitting directly across the input line, and since they are uncharged when the power supply is turned on, the ac input line sees a short-circuit condition from the time the power supply is turned on until the capacitors are charged. Inrush current limiting is necessary to prevent a very powerful surge of current, which could blow fuses and harm components in the power supply.

Inrush current limiting can be accomplished either with active components such as SCRs or passive components such as thermistors (Figure 3.9). In the SCR approach, a resistor and SCR sit in parallel in one leg of the ac line. Upon initial turn on, the SCR is off and all current flows through the resistor. As a result, the resistor limits the surge of current flowing into the input capacitors. Once the charge on the input capacitors reaches a certain critical level, a control circuit fires the SCR, thereby shorting the resistor out of the circuit and allowing normal operation of the power supply.

The SCR approach to inrush current limiting has the advantage of always resetting itself when the ac power is cycled off. Its disadvantages are more complexity due to the control circuit required to fire the SCR and slightly lower efficiency due to the constant voltage drop across the SCR while the switcher is running. The thermistor approach involves a different set of trade-offs.

A *thermistor* is a device with relatively high electrical resistance when cold and almost no resistance when at operating temperature. When the supply is first turned on, the high resistance of the thermistors prevents a large surge of current. As the input capacitors come up to full charge, the thermistors heat, losing their resistance, and allow the normal operation of the unit.

The primary disadvantage of the thermistor approach is that when temporary one- to two-cycle dropouts of the ac power occur (these happen three to four times per day on a typical ac power line), the thermistors may not cool sufficiently to limit the current when the power returns. This can cause the same set of problems as if there were no current-limiting circuitry at all. Advantages include low cost and simplicity.

The final function of the input rectifier and filter section is suppression of electromagnetic interference (EMI). The chopper section typically operates at 20 to 50 kHz and produces large amounts of electromagnetic energy. The EMI

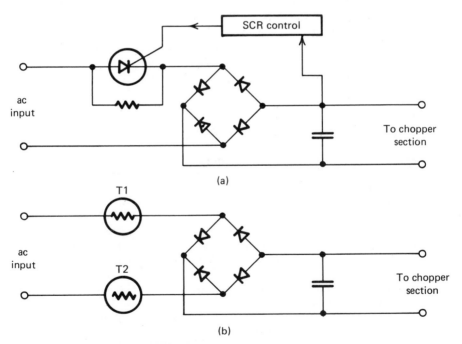

FIGURE 3.9.
Inrush current limiting can be accomplished either with (a) active components such as SCRs or (b) passive components such as thermistors.

suppression filter prevents this energy from escaping onto the ac line. The chopper section of a switching power supply is where the primary power conversion takes place. The switching transistor(s) in the chopper section converts the high dc voltage from the input capacitors into quasi-square-wave ac power.

Switching regulation can be achieved using flyback, forward, half-bridge, or full-bridge converters. Each converter technique offers a unique price–performance trade-off (Table 3.1). For example, while flyback converters are relatively simple and inexpensive, their ripple tends to become unacceptable at lower voltages and higher currents. For this reason, it is often necessary to employ one of the more complex (and expensive) converter techniques that do not produce as much ripple.

The single-transistor flyback converter is the simplest and least expensive PWM approach (Figure 3.10). Not only are flybacks inexpensive, but they are also relatively "clean" and produce less of an EMI problem, since they produce few current spikes and, therefore, less high-frequency harmonics. On the other side of the coin are the flyback's less accurate regulation, higher ripple, and relatively large transformer.

Since the transformer of a flyback is driven in only one direction, it must have an air gap and is, therefore, larger and not used as effectively as in other converter configurations. Typical flyback regulation is on the order of 1% to 1.5% compared to 0.2% to 1% for other approaches. Other consequences of the minimal use of magnetics are a higher inherent ripple and the limited voltage–current levels at which flybacks can be effectively employed.

TABLE 3.1
Switcher Price–Performance Comparison

Chopper Configuration	Advantages	Disadvantages
Flyback converter (one power transistor)	Lowest cost Very low component count	High ripple and noise Poor regulation Limited output power
Forward converter (one power transistor)	Low cost Low ripple and noise Medium output power	Inefficient use of magnetics More components than flyback
Half-bridge (two power transistors)	Efficient use of magnetics Low ripple and noise Medium power	Not economical below 100 W High component count
Full-bridge (four power transistors)	High power Efficient use of magnetics Low ripple and noise	Not economical below 750 W Highest component count

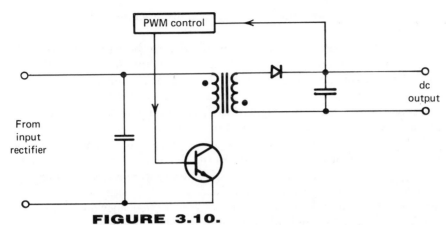

FIGURE 3.10.
Flyback converters are the simplest and least expensive switcher configuration but require a relatively large air-gapped transformer and produce less accurate regulation and higher ripple than other approaches.

Forward converters are also single-transistor converters, but they are more complex than flybacks (Figure 3.11). This configuration's flywheel diode and series connection of the inductor and load ensure that a relatively steady current flows in the transformer, regardless of the on–off state of the switching transistor. Since they use the transformer more efficiently than flybacks, forward

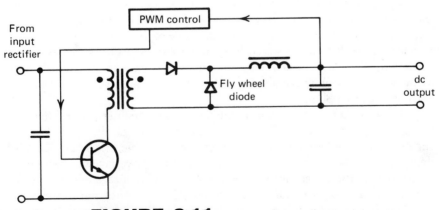

FIGURE 3.11.
Forward converters have a flywheel diode and output inductor to ensure a steady current flow through the power transformer.

converters are used more in lower-voltage and higher-current applications than flybacks.

The steady transformer current, in addition to the series inductor, combines with the forward converter to give it a lower output ripple than a flyback. The cost trade-offs in going to a forward converter are not clear-cut since it employs 50% more semiconductors (three to two) and twice the number of magnetic components, while employing a smaller output filter capacitor and a smaller power transformer as compared to a flyback.

Push-pull or bridge converters employ either two switching transistors (half-bridge) or four switching transistors (full bridge) (Figure 3.12). The transistors, or pairs of transistors in the case of full-bridge converters, are alternately switched to energize the main transformer. The advantages of this approach include higher output power, lower ripple (effective frequency doubles, making output filtering easier), and efficient, bidirectional use of the main transformer.

The negative aspects of the push-pull approach include significantly greater circuit complexity and the problem of switching transistor failure due to saturation of the main transformer core. A common cause of transformer saturation is dc imbalance caused by unequal switching characteristics of the power transistors. There are a number of circuit approaches to addressing the problem of transformer saturation, each of which adds components and further complicates push-pull converter designs.

Whichever chopper configuration is employed, if the chopper is a switcher's heart, then the control circuit is its brain. The primary function of the control circuit is to sense the output voltage, compare it to a constant reference, and adjust the duty cycle of the chopper to compensate for any changes. Most modern switchers, however, have control sections that go far beyond the basics.

It is common for the control section to include current limiting, under- and overvoltage protection, power-fail warnings, remote inhibiting of the supply, remote sensing, and a thermal switch. More sophisticated switcher control sections also include functions such as output current monitor signals, output regulation OK signals, and current sharing when operated in parallel mode. Many of today's sophisticated switchers are quasi-intelligent power subsystems, not just power supplies.

Precision references, ramp oscillators, error amplifiers, and differential voltage comparators are the basic functional elements of all switcher control circuits. All these functions and more are now available in a single IC. With their first appearance in 1976, switcher control ICs have given a great deal of momentum to the growth in switcher usage. Instead of a cage full of PC cards, today's switcher uses a handful of ICs to provide all control and housekeeping functions. This trend is expected to continue and will result in smaller, simpler, more reliable, and less expensive switching power supplies. Switchers are now a silicon-based technology and, as such, still have a long way to go before maturing.

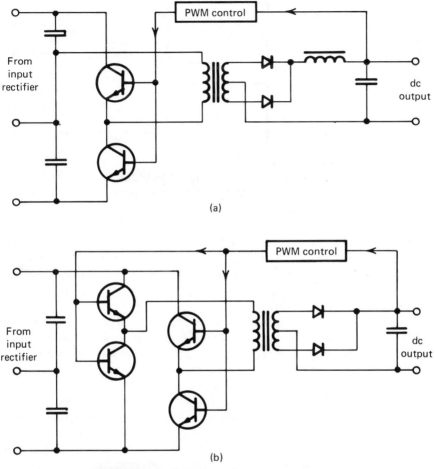

FIGURE 3.12.
In (a) half-bridge and (b) full-bridge converters, the transformer is energized in both directions, resulting in higher power-handling
capability and lower ripple.

MULTIPLE-OUTPUT
CONFIGURATIONS

Most modern electronic systems require more than a single operating voltage.
Logic is typically performed using 5 V, semiconductor memories may need 12 V,
analog circuits often operate on 15 V, and the motors in mass-storage memories,
such as floppy or Winchester disc drives, require 24 V. As a result, more and

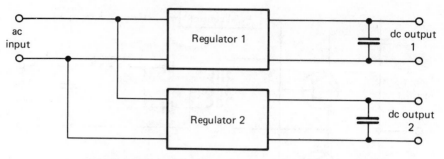

FIGURE 3.13.
The most complex multiple-output approach involves using completely independent regulators for each output.

more systems are incorporating multiple-output power supplies that can provide all required system voltages in a more compact, and possibly lower cost, package than would be possible with a separate power supply for each voltage. The internal architecture of a multiple-output supply is very important in determining its operating characteristics.

Independent regulators can be employed for each separate output (Figure 3.13). They may simply be a collection of completely separate power supplies mounted together physically, or they may actually share a common input section. In either case, it is questionable whether or not this method qualifies as a true multiple-output power supply.

The disadvantage of this approach is obviously the high parts count required for its implementation, which in itself eliminates this approach from

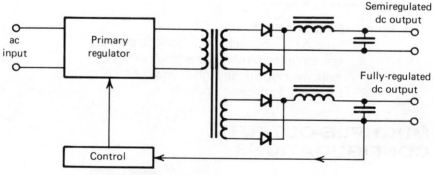

FIGURE 3.14.
This multiple-output approach is very common and gives good line and load regulation on the fully regulated output, but only line regulation on the semiregulated output.

broad usage. There are, however, a number of unique advantages that can override the high parts count in certain situations. Independent regulators yield independent outputs, they do not require minimum loading of any outputs, and they can be used to provide redundancy or multiple high-power outputs. None of these characteristics can be achieved as easily, if at all, with the two alternate multiple-output techniques.

The opposite extreme from having completely independent regulation of all outputs is to have one regulator with the secondary outputs simply derived from additional transformer windings or taps on the main winding (Figure 3.14). To employ this technique, the regulation should take place on the primary side of the power transformer. As a result, linear regulators are not generally used in this type of configuration.

The regulator provides good line regulation to all outputs, but load regulation only to the main output. As a result, the secondaries are called semiregulated, and their voltages can change as the main output loading varies. Typical secondary output load regulation is 5%. To achieve even this moderate level of secondary regulation, it is generally necessary to maintain a minimum (typically 10% of full rating) load on the main output at all times. In practical terms, this approach can only be considered when the load on one output, the main, predominates and never drops below some minimum value, such as 10%.

A middle-of-the-road technique involves a primary regulator that compensates for all line variations and provides load regulation to the main output in combination with secondary postregulators, which provide load regulation for each auxiliary (Figure 3.15). The main regulator is typically a ferro or switcher, with series pass elements used as the secondary postregulators. As in the previous approach, it is generally necessary to maintain a minimum load of 10% on the main output.

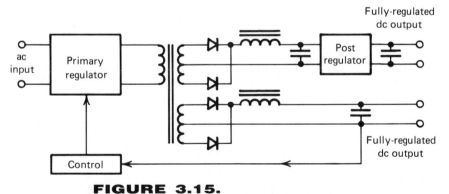

FIGURE 3.15.
Adding postregulators (usually linears) to the semiregulated outputs will yield all outputs with full line and load regulation while minimizing complexity and parts count.

Although more complex than the semiregulated multiple-output technique, this approach gives much better regulation on the secondaries without a significant loss of efficiency. The voltage drop (power dissipation) across the secondary pass elements remains at a minimum since less head room is needed as a result of the line regulation provided by the primary, nondissipative regulator. Secondary regulation can be improved from 5% in the semiregulated configuration to 0.5% or better with this fully regulated configuration.

SUMMARY

A number of circuit configurations can be used to implement the primary regulation techniques; each offers a different set of operating specifications and parameters. The regulation technique employed, and often even the specific circuit configuration, can be dictated by the power supply operating specifications required by the particular system being powered (Figure 3.16 and Table 3.2). Some of the critical parameters are overall

TABLE 3.2
Comparison of Multiple-Output
Architectures

Architecture	Advantages	Disadvantages
Separate, independent regulators	Independent outputs No minimum loading Good line and load regulation on all outputs Able to handle multiple high-power outputs	Less reliable High component count Expensive and complex
Primary regulator with semiregulated secondaries	Simplest multiple-output approach Line regulation on secondary outputs Good line and load regulation on all outputs	Poor load regulation on secondary outputs Minimum load required on primary output Typically low power secondaries
Primary regulator with postregulated secondaries	Good line and load regulation on all outputs Can have high-power secondaries All outputs are independent	Slight loss of efficiency Minimum load required on primary output More expensive than semiregulated approach

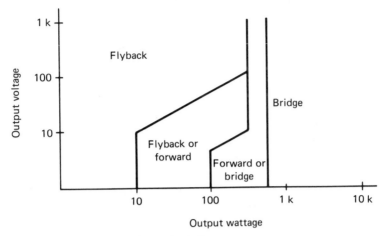

FIGURE 3.16.
Switching converter applications map.

wattage level, regulation requirements, input voltage and frequency considerations, and physical size.

At low power levels, switchers may be eliminated from consideration, while at high power levels, linears are not generally employed. Ferros are often used when bulk sources of dc current are needed for electromechanical devices such as relays and motors. In general, it is not necessary to include the desired regulation technique or circuit configuration as part of the power supply specification. If the required specifications and operating environment of the power supply are adequately defined, the available regulation techniques and even circuit configurations can be very quickly narrowed down to only one or two that are appropriate. As a result, it is necessary for power supply users to have a thorough understanding of all aspects of power supply operating parameters to ensure that they receive a unit that is truly compatible with the requirements of their system. Each regulation technique and circuit configuration fills its own specific application niche.

SPECIFICATION

4

The process of power supply selection and specification is as much a communications problem
as it is an engineering problem. A well thought
out and precisely written specification is essential if the process is to be fruitful and cost-
effective. Not only must the specifying engineer put exact desires and needs in
print, he or she must understand what the power supply vendor is saying. As will
be seen in this chapter, the meaning of power supply specifications is not always
obvious or simple.

Lack of understanding of the precise implications of particular power
supply specifications often leads to costly overspecification in an attempt to
pursue the "conservative" approach. The theory often seems to be "when in
doubt, tighten the specification (just to be sure)." A better approach would be
"when in doubt, find out the real facts and then specify accordingly." For
instance, consider the system designer who overspecifies regulation or some
other key parameter by 50%, just to be sure. Next, the power supply engineer, in
an effort to compensate for component variations, different production tech-

niques, and component aging, designs the power supply with a 50% safety factor. Given this not unrealistic scenario, it is easy to imagine a system that requires only 1% regulation for proper performance using a power supply that was specified for 0.5% regulation and actually designed to deliver 0.25% regulation.

It is easy to see that if the "conservative" approach is followed throughout the power supply specification, the result could be an extremely overspecified and more costly supply. Aside from the increased monetary cost, the power supply may be larger, heavier, and have a significantly lower reliability owing to the greater complexity that often arises out of overspecification.

Making sure the specification is complete is just as important as not overspecifying. Does this particular application require any unusual specifications such as faster-than-normal dynamic response to step load changes, or protection from high-energy spikes on the input line, or particular environmental tolerance?

PREPARING THE SPECIFICATION

The key to successful preparation of any specification is beginning with a good table of contents and a detailed outline of all the parameters to be included. It is important that the outline reflect the needs of the specific application and the abilities of the particular type of power supply, ferro, linear, or switcher, to be employed. Do not leave anything to the imagination of the vendor unless it has absolutely no bearing on the success of the system operation. The major topics included in most power supply specifications are as follows:

1. Scope

2. Applicable Documents

3. Electrical Performance

4. Environmental Requirements

5. Mechanical Requirements

6. Quality Assurance

7. Warranty and Documentation

1.0 Scope

The scope of a specification is a brief introductory paragraph that gives an overview of what type of system the power supply will be used in and any special considerations. For example, if the specifying company usually requires military-grade units but would accept a commercial-grade unit in this application, it should be noted here. Other special considerations might include such things as limited ventilation, input power source, mobile operation, redundant operation, or field replaceable.

2.0 Applicable documents

All supporting documents, such as safety and EMI standards, reliability requirements, and quality assurance manuals, that are referred to in the body of the specification should be listed here. The listing can be broken down in any number of logical orders but must be specific. For example,

> The following documents of the issue in effect on the date of this specification form a part of this control document to the extent specified herein. In the event of conflict between this specification and the reference documents, this specification shall govern.
>
> Documents
> International
> IEC 380   International Electro-Technical Commission Standard for Safety of Electrically Energized Office Machines.

The listing would go on from here and could include such documents as UL 114, VDE 0806, CSA 22.2 143, VDE 0871, FCC Docket 20780, MIL-HDBK 217, and MIL I. Headings could include International, National, Federal, Industry, Military, Safety, EMI, Quality Assurance, and others.

3.0 Electrical performance

This section begins the basics of the specification. It is here that the real differences between ferros, linears, and switchers, as well as the differences in application requirements, show up. The electrical performance section is usually subdivided into four subsections: Input, Output, Interface Signals, and General. Not all subsections apply to every situation.

3.1 Input

3.1-1 Input voltage.   Will the unit operate on 115 V ac, or on 48 V dc, or on some other input voltage? What range is it expected to vary over under normal conditions? Will the power supply be expected to accept any one of a number of possible alternatives? Ferros and switchers can generally tolerate wider input voltage ranges than can linears. It is not unusual to see a switcher specified to operate with a ±20% input voltage tolerance, whereas a linear requires ±8%, and a ferro ±15%.

3.1-2 Input frequeny.   This specification is included only for those power supplies that operate on an ac input voltage. The frequency range is typically more important than the nominal frequency, but both must be specified. Ferros require very precise frequencies for proper operation, typically 60 Hz ± 0.5 Hz. Linears are somewhat better, but they still require fairly tight input frequencies such as 60 Hz ± 3 Hz. Switchers present a mixed bag, depending on whether they are fan or convection cooled. A fan-cooled switcher is usually limited to an input frequency range of 47 to 63 Hz by virtue of the cooling fan requirements. Convection-cooled, off-line switchers, on the other hand, are limited only by the value of the input filter capacitor. A range of 47 to 440 Hz is not at all unusual for convection-cooled switchers.

3.1-3 Voltage strapping.   This specification applies only to units that can accept multiple input voltages. When this circumstance exists, it is often wise to state exactly how the conversion from one input voltage to another is to be accomplished. Will it be automatic? Will an external terminal block and jumpers be used? What about changing jumpers soldered to a printed circuit board inside the unit?

3.1-4 Input voltage transients.   Voltage transients on the input line rarely are of concern when dealing with linear- or ferro-type supplies. The large magnetics and filter capacitors in these power supplies generally will eliminate any transients that appear on the input line. Switchers are a different story. The most common failure modes in switchers involve the switching transistors themselves. In an off-line switcher, input voltage transients can come in straight off the ac line and kill the switching transistors. If the system involved is likely to operate in an environment that includes high voltage-line transients, such as an industrial or commercial workplace with electrical motors switching on and off, it may be wise to specify that the power supply be protected from input voltage transients having the following characteristics:

Time (μs)	Voltage (V)
0.0	0
0.8	1000
50.0	500
100.0	0

These characteristics are typical of the inductive spikes generated by electric motors and can be handled fairly easily using the proper transient suppressor or filter in the power supply's input filter section.

3.1-5 Input fusing If the power supply itself is to be fused, the specification must say so. Is there a preference between fuses and circuit breakers? Should the fuse be located on the front panel, on the PC board, or elsewhere?

3.1-6 Inrush current limiting This item does not usually appear on specifications for ferros or linears. It is possible for there to be a problem in a linear under certain circumstances. If, for instance, the polarity of the ac input voltage is the same at turn on as it was at turn off, the transformer is, to some degree, being double pulsed in the same direction. If this results in core saturation, the inrush current could be unacceptable. This problem rarely arises in practice.

The most common occurrence of inrush current problems arises when an off-line switcher is used. At turn on, the ac line is connected across a large, uncharged capacitor, which will act much like a dead short unless inrush current limiting is provided. Thermistor current limiting can be effective and is inexpensive. However, where ac line dropouts or system on–off–on cycling is expected to occur, the thermistors may not have time to cool enough between on cycles to provide effective current limiting. In such cases, use of thermistors can be forbidden and some type of active current limiting, such as an SCR soft circuit, should be called out (see Chapter 3).

When specifying inrush currents, both their peak levels and durations should be stated. For instance, "Inrush current is limited to 40 A peak for one-half cycle of ac." The exact values are determined by the power level of the system, the input voltage, and the amperage of the electric service to which the system will connect.

3.1-7 AC ground This is another situation that must be clearly specified. Should an ac ground termination be provided? Where and in what form of connector?

3.2 Output

3.2-1 Output voltages and currents The nominal voltages and maximum currents of all outputs should be stated.

3.2-2 Voltage adjustment range If the output voltages are to be adjustable, this section should so state. The minimum acceptable adjustment ranges should be stated, and so should where the voltage is to be measured, at the power supply terminals or at the load end of the connecting lines. In the case of high current supplies, where the voltage is measured is just as important as what the voltage is specified to be. The adjustment range itself can be called out, either as a voltage range (e.g., 4.5 to 5.5 V dc) or as a percentage of nominal (e.g., 5 V dc ± 10%).

Resolution is another often neglected aspect of the output voltage adjustment range. When left unspecified, the manufacturer has the option of using either a single-turn (poor resolution) potentiometer (pot) or a more expensive multiturn (much better resolution) pot. Many applications do not require that the voltage adjustment have a high resolution. When it is important, it must be specifically called out (e.g., "output voltage adjustment resolution shall be 5 mV or better").

3.2-3 Load current range The peak, continuous, and, in the case of multiple-output supplies, simultaneous currents for all outputs must be called out. The minimum load current should also be specifically stated. For single-output supplies, the minimum rating is not generally important; the maximum ratings are critical in properly designing the magnetics and regulating circuitry. If a multiple-output switcher is to be used, the minimum current becomes just as important as the maximum ratings. The lower the rated minimum current, the larger the output filter inductor and the slower the transient response time of the supply. Whenever possible, it is considered good practice to provide a minimum 10% load on the main output of a multiple-output switcher (see Chapter 3).

3.2-4 Line regulation This is a straightforward parameter and is usually specified as the maximum percentage of variation in the output voltage as the line voltage is varied over its specific limits. The percentage of variation may be specified as total band (e.g., 0.2%), with nominal output being the midpoint of the band, or as the variation on either side of nominal (e.g., ±0.1%). Ferroresonant supplies typically have line regulation of 1.5%, linears give excellent 0.01% regulation, and switchers occupy the middle ground at 0.1%.

3.2-5 Load regulation This sounds as simple as the previous parameter, but it is not. It should be specified as the maximum percentage of variation in the output voltage as the load current is varied from no load to full load. Be wary of specifications that define load regulation for something other than the range from no load to full load. If not specified over the full range, the regulation can be made to look better than it actually is.

A second consideration arises in the case of multiple-output supplies. Do all outputs require the same degree of regulation? Logic circuitry operates well with ±0.1% load regulation, motors can get by with ±5% or less regulation, and sensitive analog circuits may require 0.01% regulation.

Ferroresonant supplies cannot provide tight load regulation without linear postregulators; switchers can provide ±0.1% but cannot be used with sensitive analog circuits. Linears can provide the precise regulation demanded by even the most particular analog circuits. Most of today's sytems are digital and do not require extremely precise load regulation; as a result, switchers can meet the requirements for load regulation of the vast majority of systems.

3.2-6 Interaction This specification applies to semiregulated multiple-output supplies. The interaction between two outputs is the percentage of variation in the output voltage of one output while the other output is varied from no load to full load. This can be an important parameter in many system applications. The worst interaction is usually related to changes in the auxiliary output voltages as a result of changes in the load on the primary output. Supplies that do not have postregulated secondaries can have interactions of 2% and greater. Postregulated secondary outputs usually have an interaction of no more than the percent of variation specified for individual load regulation (e.g., if load regulation is ±0.1%, then interaction will be ±0.1% or less). Supplies in which the main output is fully regulated and the auxiliaries are unregulated are generally used only when the main output predominates and is not subject to wide variations in output loading.

3.2-7 Centering Centering is the variation of an output from its nominal voltage caused by design limitations and manufacturing tolerances. Centering is expressed as a percentage ratio of volts variation to volts nominal. It is usually measured with all outputs at 50% of maximum load and with the primary output adjusted to nominal.

Centering is only specified for semiregulated multiple-output power supplies. In most cases, centering error is primarily due to the necessity of having an integral number of turns on each winding of the power transformer. Centering error will generally be 5% or less.

3.2-8 Differential-mode PARD This is what is typically referred to as noise and ripple, especially if only a single specification is given for noise and ripple. Linears and ferros tend to be very quiet, with typical specifications calling for 0.01%. It is important that this be called out as a peak-to-peak value. Switchers are inherently noisier than other power supplies. The ripple and noise produced by switchers is characterized by high-frequency spikes. As a result, the rms value may be quite low, 5 mV or less, while the peak value may exceed 100 mV. A typical switcher will have a maximum of 1% or 50 mV of peak-to-peak ripple and noise on a 5-V output. It is also good practice to specify the bandwidth over which PARD is to be measured. Typically, the bandwidth will run from dc to 30 megahertz (MHz).

3.2-9 Common-mode PARD This aspect of noise and ripple is commonly ignored by power supply specifications. One reason is that it is almost a nonentity as far as linears and ferros are concerned and, even in switchers, rarely causes system problems. If it is to be limited, it should be treated similarly to its differential-mode cousin. Typical levels of common-mode PARD in switchers are 4% or 200 mV for a 5-V output, measured across an impedance of 10 ohms

(Ω) over a bandwidth from dc to 30 MHz. Notice the addition of the 10-Ω impedance, which does not appear in the differential-mode version.

3.2-10 Overshoot and undershoot These parameters are specified under two conditions, the first being turn on and turn off. A typical statement would be, "There shall be no overshoot or undershoot upon turn on or turn off." This applies to all types of supplies. The other condition where these parameters come into play is step function changes in load current. For step changes in load, ferros, linears, and switchers all react differently. Linears will have the least over- and undershoot, typically less than 1%, switchers will be in the 1% to 2% range, and ferros will often be over 5%. When specifying the maximum deviation, it is important to state the amplitude and rate of the step load change. For example, "Maximum output voltage deviation shall be 2% with a step load change of 25% at 5 A per microsecond."

3.2-11 Response Time How quickly the output returns to stay within the regulation band after a step load change is the *response time.* Since a step load change is involved, this specification also varies from linears to switchers to ferros. Response time depends on the amount of dampening in the feedback loop of the power supply and the bandwidth. Linears typically have the ability to respond in a few microseconds to a step load change, a 20-KHz switcher will take a few hundred microseconds, and a 60-Hz ferro will take a few thousand microseconds. Response time is like regulation in that it is often overspecified. For most applications the mid-ground offered by a switcher is more than adequate. A typical specification might read, "Response time shall be a maximum of 200 microseconds to 1% after a 25% step load change at 3 A per microsecond."

3.2-12 Temperature coefficient A power supply's *thermal regulation* is its temperature coefficient. It is expressed as the percentage of change in output voltage per degree Celsius of change in ambient temperature. A typical specification might read, "The temperature coefficients of all outputs are 0.02%/°C maximum." The value is chiefly determined by the temperature characteristics of the components used and, to a lesser extent, by circuit design. In most applications, the ambient temperature will not be expected to vary too much once the initial warm-up period is over, and a coefficient of 0.02%/°C is more than adequate. In those few instances where operating ambients are expected to vary widely, temperature regulation can become an issue. Consider a 30°C change in ambient with a 0.02%/°C temperature coefficient. The output voltage could change 0.6% just owing to temperature factors. Add that to line and load effects, and a possible step load change, and the total regulation of the output can deteriorate to an unacceptable level quite rapidly. It must be noted that, while possible, this scenario is very unlikely since it depends on an unusually wide change in ambient operating temperature. Ferros, linears, and switchers all have similar temperature coefficients.

3.2-13 Holdup time Specifications for linears and ferros almost never include this parameter since, for all practical purposes, they have no useful holdup time. Holdup time is the time period during which the output remains in regulation after removal of ac input power. It is a significant benefit offered by switchers and is almost solely dependent on the value of the input filter capacitors, operating efficiency, and the load.

Typical designs will yield 16 milliseconds (ms) (one cycle at 60 Hz) to 50 ms (three cycles) of holdup time. Holdup time should always be specified at full load and either nominal or low line. The holdup time for a given switcher varies inversely with output load and directly with input voltage.

3.2-14 Turn-on delay Turn-on delay is the elapsed time between the application of input power and the attainment of all output voltages of their nominal values. It is typically specified not to exceed 1.0 second(s). This specification applies primarily to multiple-output units and is the same for ferros, linears, and switchers.

3.2-15 Load capacitance or inductance This parameter defines the load that the power supply "sees" in terms of capacitance or inductance, as applies to the particular system. In typical digital systems with their decoupling capacitors, this specification might read, "All outputs shall exhibit stable performance under all specified operating conditions when the remote sensing location is loaded by a capacitance of 3 to 8 microfarads (μF) per load ampere." It is also possible to have some outputs supplying capacitive logic circuits and others supplying inductive circuits, such as disc drives or printer servos; the specification should be stated separately for each output if necessary. It does not vary from ferros to linears to switchers.

3.2-16 Remote sense Compensation for power distribution cable drops is often provided on medium- to high-power outputs and is called remote sense. It essentially moves the point of regulation from the power supply end of the cable to the load end of the cable. Generally found on switchers or linears, it is specified as "remote sense shall compensate for up to a 500-mV drop in the power distribution cables." Typical values range from 200 to 500 mV.

There are less obvious considerations when specifying remote sense. What will happen if one or both sense leads are lost? What if the sense leads are reversed accidentally? To cover these possibilities, a clause can be added such as "Loss of sense leads will not cause the respective output to change more than 5%. Reversal of sense leads will not cause catastrophic failure of the power supply."

Other concerns related to remote sensing include the possibility of the power supply going into oscillation if remote-sensing a load with continuous or rapid step load changes, and the fact that the real output power of the supply includes both the power dissipated by the load and the power dissipated in the power distribution cables (see Chapter 7).

3.2-17 Output polarity Here is another parameter that relates solely to multiple-output units and applies equally to ferros, linears, and switchers. If the output voltages of paragraph 3.2.1 were called out with specific polarities such as V1 = +5 V, V2 = +12 V, and so on, it will be necessary to state whether all or only certain groups of outputs share a common ground line. In the case of independent outputs, a statement should be included about how far off a chassis ground each output may be referenced—for example, "All outputs are independent and may be referenced as desired up to 100 volts off chassis ground." The particular voltage limitation results from the voltage ratings of the output decoupling capacitors. In some instances, there will be both polarized and floating outputs in the same unit. In that case, it is especially important to state how the ouptuts are referenced and which, if any, share a common ground.

3.2-18 Overvoltage protection Overvoltage protection is normally specified for linears and switchers and occasionally specified on ferros. A common failure mode of linears is an emitter–collector short in the pass transistor, which causes unregulated, higher voltage dc to appear at the output terminals. In switchers, it is possible for the output voltage to rise owing to an open circuit in the feedback loop, but it is somewhat unusual. In either case, load protection against these higher voltages is provided by the overvoltage protection (OVP) circuit.

In linear supplies, OVP is almost always provided by an SCR crowbar circuit that shorts the high voltage to ground. Switchers either can employ an SCR crowbar or may have electronic limiting built into the PWM control circuit. The statement should be specific in stating the type of OVP required, since an SCR crowbar has a quicker reaction time but will generally be more costly.

Typical OVP circuits are not automatically reset until the input power is cycled off. The reason for the latching nature of most OVP circuits is that the tripping of the OVP generally indicates a fault condition somewhere in the power supply. The OVP is meant to protect the load and latches the supply off until the supply is cycled off (the fault condition is eliminated), when the supply is cycled on again (see Chapter 7).

3.2-19 Stability This paragraph states the maximum allowable percent of change in output voltage over a specific time period, typically 24 hours, once thermal equilibrium has been reached. The time allowed for warm-up is also stated, typically 30 minutes to 18 hours, at which time thermal equilibrium is assumed to exist. The stability parameter is also referred to as *drift*.

3.2-20 Sequencing If power-up or power-down voltage sequencing is required, the sequence time delay and order of output voltage appearance must be stated.

3.2-21 Margining Margining is used to change the output voltage by a small percentage (typically ±5%); it is generally used as a system testing tool. Weak or out of tolerance circuits can often be detected by using this feature. It is better to force the fault condition rather than wait for some combination of line regulation, load regulation, temperature change, and the like, to cause a change in output voltage that results in a random fault in system operation.

The percent of variation required in output voltage should be stated, as well as whether the margining will be locally or remotely implemented. Local margining requires that the power supply manufacturer include a switch (typically SPDT, center-off) on the front of the supply. Remote margining requires that a set of output pins be provided to which the system manufacturer can wire his own switch. This specification applies to all types of supplies.

3.2-22 Programming This can be a confusing term since it is used, in some instances, to refer to output sequencing and, at other times, refers to output margining. In general, the terms "programming" and "remote programming" should not be used. Instead, the more descriptive "sequencing" and "margining" should be used. Properly used, programming applies to laboratory-type power supplies that can be controlled to deliver a series of different voltages or currents on the same output.

3.3 Interface signals

This section contains a series of paragraphs describing all the required interface signals. These signals are often incorporated in switchers and include such things as power fail detection, remote inhibit, output current monitor, remote margining, dc OK, and OVP-actuated signals. Only two fairly standard signals are described here, since other signals are typically tailored to each specific application.

3.3-1 Power-fail detection Used exclusively in switchers, this signal provides early warning to the host system that the input power is gone and, within some specified time, the output voltage of the supply will drop out of regulation. This signal allows the system to make use of the holdup time for "housekeeping," such as saving data or software prior to loss of power. A typical statement would be, "Upon loss of input power, a TTL compatible signal (logic 0) is provided, which is referenced to the primary output negative sense terminal. This signal appears at least 5 milliseconds prior to the output dropping 5% below its nominal value." Other pertinent information might include the current sinking and sourcing limits of the circuit and possibly a timing diagram showing the dynamic relationship between loss of input power, the power-fail signal, and loss of output regulation (see Chapter 7).

3.3-2 Remote inhibit Also called remote on–off or logic inhibit, this signal is an input to the power supply and controls the output voltage. A typical statement would be, "This supply can be remotely turned on with an open circuit (logic 1) and turned off with a switch closure (logic 0) referenced to the primary output negative sense terminal."

3.4 General

3.4-1 Overload protection Also known as current limit or power limit, this circuit protects the power supply from externally induced overloads. It is an inherent characteristic of ferros and can be designed into linears and switchers. Two common types of current limiting are constant current limiting and fold-back current limiting. The type of protection should be described and the following statement added: "Short-circuit operation can be continuous without damage. Recovery is automatic when overload is removed" (Chapters 6 and 7).

3.4-2 Efficiency The required level of operating efficiency should be explicitly stated, along with the specific operating parameters under which it is to be measured, such as "Minimum efficiency shall be 70% at full load and nominal ac line." Efficiency generally declines at less than full loading and varies as the ac line voltage changes. The line- and load-related variations in the efficiency of linears are much more pronounced than they are in ferros or switchers.

Typical efficiency levels are 70% for switchers and 80% for ferros. The typical efficiencies of linears are related to the output voltage level: low-voltage units (e.g., 5 V) will be about 40% to 50% efficient; high-voltage units (e.g., 28 V) will be 50% to 60% efficient.

3.4-3 Thermal protection Protection from excessive heat is often built into power supplies, typically in the form of a thermal switch. It is common in all types of supplies and is generally automatically reset when the excessive temperature is eliminated.

3.4-4 No-load protection Most power supplies can be operated with no load on the output without damage. This specification typically states that the power supply will not suffer reduced reliability or otherwise impaired performance when proper loads are reapplied.

3.4-5 Parallel operation It is not uncommon for power supplies to be operated in a parallel–redundant mode in many of today's sophisticated systems. When supplies are to be operated in parallel, will it be a master–slave configuration or straight paralleling? What about current sharing and inclusion of blocking diodes? All these specifics must be included (see Chapter 7).

3.4-6 Safety If the supply is to meet any specific safety standards, such as IEC or UL, it should be stated here, referring to the documents cited in paragraph 2.0.

3.4-7 Electromagnetic interference Similar to paragraph 3.4-6 if the supply is to meet any specific EMI standards.

4.0 Environmental requirements

This section defines the environmental parameters in which the power supply will "live." This includes storage conditions as well as operating conditions. The parameters defined in the environmental requirements do not vary from ferro to linear to switcher; they are dependent on the when, where, and how of the operation of the final system.

4.1 Storage temperature

Generally a wider range of temperatures than the operating temperature range, the storage temperature is dependent on the quality level of the components used (e.g., military specification components versus industrial-grade components). A typical specification would read, "Storage temperature shall be $-55°C$ to $+85°C$."

4.2 Operating temperature

The range of temperatures that the final system will function in is the operating temperature range. The operating temperature range should clearly state that full output power is to be maintained between the temperature limits. If that is not done, the power supply may indeed operate at the high end of the range, only with a severely derated (reduced) output power.

A typical specification allowing derating is, "The operating temperature range is $0°C$ to $+71°C$. Full power is maintained to $50°C$, with the output power linearly derated to 60% of nominal at $70°C$." This is quite different from a power supply that can deliver full rated power over the entire range.

4.3 Relative humidity

The relative humidity is not critical for most low-voltage applications, but it can become very important if high voltages are used in the system. Under high voltage, different insulators and connectors may all be required. A typical specification is, "The relative humidity range shall be 5% to 80%, noncondensing."

5.0 Mechanical requirements

The mechanical configuration of a power supply is often neglected, but it is equally as important as the electrical specifications. Key aspects of power supply mechanical requirements include size, outline shape, cooling, connectors, mounting, and weight. Some of these parameters, such as weight and overall size, can be significantly affected by whether the electrical specifications call out a ferro, linear, or switcher. The other parameters, such as cooling, connectors, and mounting, are more independent of the power supply technology used.

5.1 Mechanical outline

The overall size and shape should be described in writing with reference to an outline drawing included as an integral part of the specification. The mechanical outline drawing itself should include all mechanical information, such as overall exterior dimensions, size and placement of mounting holes, types and locations of all electrical interface connectors, and airflow pattern.

5.2 Mounting holes

The size (e.g., 8–32 unc) and location of all mounting holes should be called out, with reference made to the mechanical outline drawing.

5.3 Cooling

The type of cooling (convection or forced air), as well as airflow direction, should be stated, with reference made to the mechanical outline drawing. Other considerations or requirements are as follows: Should the airflow on a forced-air unit be unidirectional (nonreversible) or bidirectional (reversible)? Will convection-cooled units be mounted in such a fashion that the system will provide additional heat sinking or airflow? Power supply cooling is a critical parameter determining operating life and performance. The cooling specification must be approached in a well thought out and detailed manner to ensure proper supply operation (Chapter 5).

5.4 Weight

Maximum weight should be stated.

5.5 Connectors

Both the types of connectors required and their associated pinout locations must be specified.

5.5-1 Connector types The specific connectors required for the input, outputs, and interface must be clearly stated. Are plug type, card edge, barrier strips, or some other connector style required? When appropriate, a specific vendor's part number should be included. Each required connector and its function must be included.

5.5-2 Pinout layout Complete details of all signal and power pinouts for each connector must be provided. This is generally done in a listing format, such as the following:

	J1	J2
Pin 1	+ sense V1	power fail inverse
Pin 2	− sense V1	current monitor
Pin 3	remote inhibit	+ sense V2
Pin 4	power fail	− sense V2
Pin 5	high margin	N.C.
Pin 6	low margin	− sense V1

5.6 Markings

Markings include identification of all connections (e.g., +V1 or GND), ratings (e.g., V1 = 5 V at 100 A), manufacturer's or user's identification number, date codes, and the like. If position or form of the markings is important, it must be stated. It is good to also state, "Markings shall remain legible throughout the normal operating life of the power supply."

5.7 Shock and vibration

If the power supply is expected to meet any shock and vibration specifications, they should be explicitly called out here, or reference should be made to another controlling document, which itself is called out in paragraph 2.0. Even if the final system is operated in a stationary environment, the power supply can be subjected to substantial shock and vibration during shipment both inside and outside the system and during installation in the system.

6.0 Quality assurance provisions

This section includes all inspection, quality assurance, and testing requirements. Sections for initial product qualification testing and for ongoing product testing may both be included. It is also possible to call out specific testing procedures in addition to the specific parameters to be tested. The initial qualification testing is generally much more thorough (including *all* specified parameters), while the ongoing test program includes only critical parameters such as output voltage, output current, line regulation, and load regulation, as required by the host system. Also included here are burn-in requirements, documentation of testing, and any design reliability (MTBF) calculations that must be supplied (Chapters 5 and 6).

7.0 Warranty and documentation

This section details all the loose ends involved in finalization of the documentation and ensuring adequate warranty coverage after the power supply is delivered. The warranty period and how warranty service is to be implemented should be stated. Will warranty service be performed on site (very unusual in the case of power supplies), or will the unit be non-field repairable (typical of switchers)? Are parts lists and suggested vendors needed? What about schematics or operation manuals and the like?

SUMMARY

Specifying electronic systems or components is never easy, and power supplies are no exception. Knowing how to write a power supply specification can be a great help when reviewing specifications submitted by potential vendors. On the surface, it appears rather easy to write a specification that describes the required power supply. However, if you are not careful, you may in fact detail a power supply that cannot be economically produced or, worse yet, will not work in your system.

These problems can arise from overspecification, resulting in a power supply more complex and expensive than required, and from overlooking a seldom specified characteristic, or specifying it ambiguously, resulting in a supply that does not function properly in your system.

Much has been said about the relative advantages of switchers. They are smaller, lighter, and more efficient than linears or ferros. One of the trade-offs made to realize the benefits of switching power supplies is the much greater complexity of their circuitry and specifications. For the system designer, it is no longer enough to order "5 V at 50 A"; he or she must purchase from a well thought out and detailed specification.

RELIABILITY

Reliability is important to understand and quantify both because it can directly impact the operating success of a design and because it costs money. Because reliability does cost money, a trade-off exists between the costs of increasing reliability and the costs resulting from system failure after delivery (Figure 5.1).

Before the trade-off between cost and reliability can be addressed, both must be quantified. This chapter will focus on the techniques commonly used to predict power supply reliability and on the thermal considerations that most directly affect reliability.

MEAN TIME
BETWEEN FAILURES

Reliability is a key consideration in the design of any electronic system. Depending upon the circumstances, system downtime can be inconvenient, costly, or

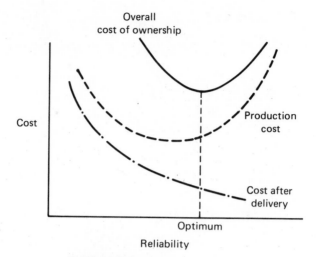

FIGURE 5.1.
Reliability versus cost trade-off.

even disastrous. Power supplies (as well as all electronic systems) go through three distinct periods of varying reliability (Figure 5.2).

The period directly following initial turn on is called the *burn-in* time and is characterized by a relatively high failure rate. The failure rate, λ, is defined as the ratio of the total number of failures to the total hours of operation for a given population.

$$\lambda = \frac{\text{failures}}{\text{hours of operation}} \tag{5.1}$$

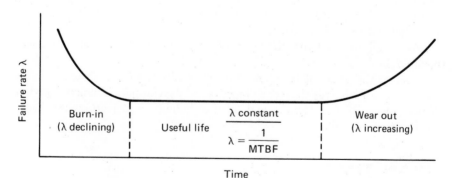

FIGURE 5.2.
"Bath-tub" curve of product life cycle reliability.

The high initial failure rate can result from weak components, excessive tolerance buildups, poor workmanship, design problems, and so on. A second important characteristic of the burn-in time is the rapidly declining rate of failures. In the case of power supplies, most early failures are experienced within about 12 hours of initial turn on.

Following the burn-in period is the *useful life* of a power supply. During this time, the failure rate is low and relatively constant. At the end of a power supply's useful life, the failure rate begins to rise. This is the *wear-out period* and results, as the name implies, from the wearing out or degradation of the components.

In practical terms, the failure rate of a power supply during its useful life (i.e., after adequate burn in and before wear out) is the most important reliability parameter. During the useful lifetime, only random failures should appear. It is these chance failures with which most reliability studies are concerned.

The parameter commonly used to compare the reliabilities of power supplies is the *mean time between failures* (MTBF). The MTBF for a given design is defined as the inverse of the failure rate during its useful life:

$$MTBF = (\lambda)^{-1} \qquad (5.2)$$

The true failure rate, and therefore the true MTBF, for any given design cannot be calculated until after the last unit has been taken out of operation. In the real world, there is little point in waiting until the end of the useful life of a power supply design to calculate MTBF. What is needed are accurate, comparable, and timely estimates of MTBF so rational decisions can be made at the design stage.

It is important to note that the MTBF of a power supply is a statistical representation of the *expected* mean time between failures. In general, it is not a measured quantity but a prediction based on a priori data. While the MTBF of a power supply is important, it in no way represents the *minimum* time between failures.

SAMPLE METHOD OF MTBF PREDICTION

Statistical sampling methods can be used to relate the observed failure rate of a representative group to that of the whole population with a given level of confidence:

$$MTBF = \frac{1}{1 - (1 - C)^{1/T}} \qquad (5.3)$$

where C is the level of confidence and T is the total operating time without failure (or the total operating time until just prior to the first failure in the sample group).

Consider twenty 500-W power supplies run for a period of 3000 hours, each without any failures. If the MTBF is desired with a 60% level of confidence, the estimated value in the case would be an MTBF in excess of 65,000 hours.

Common practice is to use confidence levels of either 60% or 90% when predicting MTBF. Most power supply MTBFs are calculated at the 60% level of confidence, since it yields a higher predicted value. It is important to know the

A large number of variables are involved in the operation of any sample group of power supplies. Were the supplies run at high, low, or nominal input voltage? What was the ambient temperature? Did the temperature cycle from high to low and back? Were the supplies run continuously or periodically cycled on and off? What about vibration, humidity, load factors, and so on?

Calculated MTBFs derived from the sample method are very difficult to use when comparing power supplies, owing to the many variables involved. This is especially true when different manufacturers become involved.

The second problem with the sample method is that, even though the MTBF can be estimated long before the end of the useful life of a power supply, it still requires that several units be tied up in testing for an extended period of time, over 4 months in the earlier example. Therefore, although the sample method can provide good estimates of MTBF, it is not generally used in the power supply industry.

COMPONENT FAILURE RATE MTBF PREDICTION

Because of the inherent difficulties with the sample method, the component failure rate (or parts stress analysis) method is used by most power supply manufacturers to predict MTBF. This method calculates the MTBF as a function of the failure rates for the component parts used to manufacture the power supply:

$$\text{MTBF} = \frac{1}{\sum_{i=1}^{n} \lambda i} \tag{5.4}$$

where n is the number of components and λi is the expected failure rate of the ith component.

The key variables in this method are the component failure rates. These failure rates are provided in MIL-HDBK 217 and are determined using a component failure rate model based on

$$\lambda_{ip} = \lambda_{ib}(\pi_{i1} \cdot \pi_{i2} \cdot \pi_{i3} \cdot \ldots \cdot \pi_{in}) \tag{5.5}$$

where λ_{ip} is the component failure rate of the ith component, λ_{ib} is the base failure rate, and the π_i's are the environmental and other parameters that modify the base failure rate and affect component reliability (Table 5.1).

TABLE 5.1

Representative π Factors from MIL-HDBK 217

Factor	Description
Common to all component categories	
π_E	Environment: accounts for influence of environmental factors other than temperature
π_Q	Quality: accounts for effects of different component quality levels
Discrete semiconductors	
π_A	Application: accounts for effect of application in terms of circuit function
π_C	Complexity: accounts for effect of multiple devices in a single package
π_{S2}	Voltage stress: adjusts model for a second electrical stress in addition to wattage
Resistors	
π_R	Resistance: adjusts model for the effect of resistor ohmic values
π_C	Construction class: accounts for influence of construction of variable resistors
Capacitors	
π_{SR}	Series resistance: adjusts model for the effect of series resistance of electrolytics
π_{CV}	Capacitance value: adjusts model for the effect of capacitance related to case size

The component failure rate method of predicting MTBFs provides a more direct way to compare the reliability of various power supplies. However, even this method of comparison must be used with care. The precise method used to calculate MTBF can vary from one manufacturer to another. Some manufacturers include only the "necessary" components in their calculations (excluding components in the on–off, power fail, margining, and other "nonessential" circuits). This procedure of excluding selected components tends to bias the calculated MTBF upward and can cause serious difficulties in comparing the reliability claims of various manufacturers.

Predictions of MTBF based on the component failure rate model find that linears appear to be about 1.5 times more reliable than equivalent switchers owing to a lower parts count (Table 5.2). In addition, linears have no high-frequency switching circuitry to generate high-voltage transients. The calculated MTBF for a typical power supply can, however, be a very misleading figure.

One particular commercial 750-W switcher has a calculated MTBF of only 18,000 hours but has achieved a documented MTBF of 200,000 hours in the field! Such divergence between the actual and calculated MTBFs for switchers

TABLE 5.2

Comparison of Linear and Switcher
MTBFs per MIL-HDBK 217

Component	Failure Rate (λ) Com. Parts	Mil. Parts	Linear Qty. N	Com. λN	Mil. λN	Switcher Qty. N	Com. λN	Mil. λN
Capacitors								
Ceramic	0.066	0.022	2	0.132	0.044	30	1.98	0.66
Film	0.078	0.026	1	0.078	0.026	3	0.234	0.078
Tantalum	0.2	0.063	1	0.2	0.063	7	1.4	0.441
Electrolytic	1.6	0.48	7	11.2	3.36	5	8.0	2.4
Resistors								
Fixed, composition	0.0045	0.0045	14	0.063	0.063	90	0.405	0.405
Fixed, wire-wound	0.1	0.1	22	2.2	2.2	5	0.5	0.5
Fixed, film	0.003	0.003	4	0.012	0.012	6	0.018	0.018
Var., cermet	2.76	1.38	2	5.52	2.76	4	11.04	5.52
Var., wire-wound	0.3	0.1	2	0.6	0.2	2	0.6	0.2
Diodes								
Signal	0.028	0.0056	6	0.168	0.034	55	1.54	0.308
Power	2.16	0.432	2	4.32	0.864	7	15.12	3.024
Transistors								
Signal, NPN	0.071	0.014	4	0.284	0.056	9	0.639	0.126
Signal, PNP	0.1	0.02	2	0.2	0.04	8	0.08	0.16
Power, NPN	0.65	0.13	12	7.8	1.56	4	2.6	0.52
Thyristors								
Power	1.8	0.36	φ	φ	φ	2	3.6	0.72
ICs, digital								
≤20 gates	1.05	0.018	φ	φ	φ	1	1.05	0.018
ICs, linear								
≤32 Qs	0.9	0.033	2	1.8	0.066	4	3.6	0.132
Transformers								
Signal	0.01	0.003	1	0.01	0.003	3	0.03	0.009
Power	0.082	0.032	2	0.164	0.064	3	0.246	0.096
Switches								
Thermal	0.225	0.003	1	0.225	0.003	1	0.225	0.003
Connectors								
Contact pairs	0.032	.004	16	0.512	0.064	25	0.8	0.1
Fuses	0.3	0.1	1	0.3	0.1	2	0.6	0.2
Totals			104	35.8	11.6	276	55.0	15.6
Calculated MTBF (hours)				28,000	86,000		18,000	64,000

comes primarily from two sources. First, adding components that enhance actual reliability (clamping diode, snubbers, etc.) tends to lower the calculated MTBF. Second, because switchers are more efficient, they dissipate less heat, and less heat means lower stress and longer component life. The calculated MTBF can be a useful tool for comparing one switcher to another or one linear to another, but it loses much of its meaning when used to compare a linear with a switcher.

GENERAL THERMAL CONSIDERATIONS

Heat is the primary enemy of electronic circuitry, and power supplies are no exception. Improved cooling is the single most important step that can be taken toward improved power supply reliability.

Heat sources in electronic systems generally result from current flowing through some electrical resistance. In power supplies, the main sources of heat are transistors, diodes, resistors, and transformers.

The transistors are the most fragile heat-generating components in a power supply. In linears, they are used to dissipate relatively large amounts of power and, therefore, heat. In switchers, they experience less heat, but they are subjected to much higher voltage stresses.

The expected failure rate of silicon transistors rises rapidly at junction temperatures in excess of 125°C. An increase from 125°C to 150°C will approximately double the failure rate. For that reason, the case temperature should not be allowed to rise above 110°C for silicon devices and 90°C for germanium and Schottky devices.

The semiconductor case temperature (switching transistors and output diodes in switchers and pass transistors in linears) is the single most effective and practical estimate of overall power supply operating stress.

In switchers, the output diodes are usually a major source of heat, which results from the current flowing through the forward voltage drop and from the reverse recovery current flowing during the recovery period. The use of Schottky output diodes can significantly reduce the generated heat due to their lower forward voltage drop. Reverse recovery heating, although significant in switchers, is negligible at the 60-Hz operating frequency of most linears.

Some of the resistors used in switchers, usually in bleeders and snubbers, can dissipate a relatively large amount of heat. Fortunately, though, resistors are very rugged and do not often require special cooling.

Transformers can also contribute to power supply heating owing to the IR drops in their windings and because of the molecular action within the core caused by the changing magnetic flux. Long-term high-temperature operation can degrade potting compounds and enamel wire insulation, causing failures due to turn-to-turn shorts in the windings. Fortunately, this is rare and usually occurs only when a power supply is operated in a harsh environment for extended periods.

THERMAL CALCULATIONS: CONVECTION COOLING

Given the importance of power supply operating temperature to reliability and the many sources of heat generation in a power supply, it becomes apparent that an important aspect of power supply selection is the determination that the supply will perform to the rated specifications within the actual system environment. A good understanding of the basic principles involved in heat transfer is necessary when making that determination.

The output power of a dc output power supply is simply the product of the output voltage and the output current and is measured in watts. The input power of a power supply is also the product of voltage and current, but it must be averaged over a full power input cycle of ac input supplies (see also Chapter 6):

$$P_{in} = P_{avg} = \frac{1}{T} \int_0^T V(t) i(t) \, dt \tag{5.6}$$

Power supply efficiency is defined as the ratio of output to input power expressed as a percent:

$$\eta = \frac{P_{out}}{P_{in}} \times 100\% \tag{5.7}$$

As discussed earlier, there are many sources of power loss caused by the components within a power supply. The result is the case temperature rise of an operating power supply. This dissipated energy or lost power can be expressed as

$$P_{loss} = P_{in} - P_{out} \tag{5.8}$$

By combining Equations (5.7) and (5.8), internal power loss can be expressed as a function of efficiency and output power:

$$P_{loss} = P_{out} \left[\frac{1 - \eta}{\eta} \right] \tag{5.9}$$

It is also possible to express the power (heat) dissipated per square unit of surface area (watts per square inch) as

$$\phi_{dis} = \frac{P_{loss}}{S} \tag{5.10}$$

where S is the surface area that dissipates the internally generated heat from the source to the surrounding environment.

Curve I of Figure 5.3 represents the Stefan–Boltzmann law and plots the surface temperature rise of an enclosure as a function of $\emptyset_{dis}$, assuming the thermal emissivity (conductivity) of the surface of the enclosure is 0.9. The Stefan–Boltzmann law describes the rate of emission of radiant energy from the surface of a body by

$$R = e\sigma (T_2 - T_1) \qquad (5.11)$$

where e is the emissivity, T_2 and T_1 are the absolute temperatures of the radiating surface and the ambient, respectively, and σ is a constant.

Again referring to Figure 5.3, note that curve II, a straight line with a slope of 125°C/W/in.², closely approximates the Stefan–Boltzmann law as indicated by curve I. Therefore, given the output power, efficiency, and geometry of a power supply, calculating the average case temperature rise becomes a simple matter (assuming a surface emissivity of 0.9).

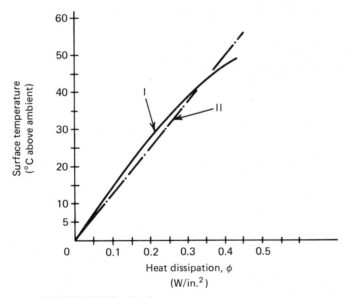

FIGURE 5.3.
Stefan–Boltzmann law: curve I represents the behavior of a radiating black surface with an emissivity of 0.9; curve II is a straight-line approximation with a slope of 125°C/W/in.².

The following procedure can now be used to estimate the average case temperature rise ΔT_c of a power supply with radiating surface S, output power P_{out}, and efficiency η:

1. Determine the radiating surface area S.

2. Solve $P_{loss} = P_{out} \left[\dfrac{1 - \eta}{\eta} \right]$.

3. Solve $\phi_{dis} = \dfrac{P_{loss}}{S}$.

4. Solve $\Delta T_c = \dfrac{125°C}{W/in.^2} \times \phi_{dis}$ \hfill (5.12)

Whether you design power supplies in-house or buy them from a vendor, a knowledge of the expected average case temperature rise is important to have. The laws of thermodynamics and electronics determine the overall size of a power supply. Without forced-air cooling, the radiating surface area and the operating efficiency will determine a power supply's average case temperature rise for any specified output level.

IMPROVED
CONVECTION COOLING

In most convection-cooled power supplies, the internally generated heat is conducted to the outside world through a finned aluminum extrusion. These extrusions are commonly called *heat sinks,* but they are actually only pathways by which heat travels to the air mass around the power supply. It is the air mass that is the ultimate heat sink.

As shown earlier, the emissivity is an important parameter in determining the rate at which heat is conducted away from a given heat sink [Equation (5.11)]. Given the physical packaging constraints of any particular power supply design, what can be done to improve emissivity and thereby cooling?

First, the heat sink can be black anodized instead of the more common silver color. Black anodization can increase effective heat transfer by about 8%. Conventional heat sinks, however, which have parallel, straight, smooth fin surfaces, suffer from important limitations, whatever their color.

Heat transfer to the air is greatest when the airflow is turbulent rather than streamlined or laminar. Placed vertically, the fins in conventional heat sinks cause the free convection airflow to be laminar, thus limiting heat transfer. When placed horizontally, these parallel fins impede the heat's natural upward flow; that is, heat transferred to the air by the lower fins can end up flowing back into the upper fins.

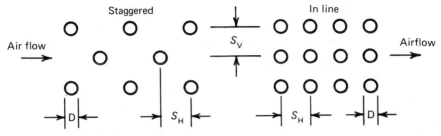

FIGURE 5.4.
Pin-fin geometry. Staggering the pins produces more turbulence in the convection
airflow and results in improved cooling compared with an in-line pin geometry.

Pin fins (Figure 5.4), however, provide both increased surface area and better flow characteristics than extruded fins. They allow convection flow in almost any mounting orientation, and they tend to produce a turbulent, rather than laminar, airflow.

In general, the more pin rows in the direction of the airflow, the greater the turbulence and the higher the rate of heat transfer. The ratio of the spacing between pins to the pin diameter also affects the results. Optimum ratios of S_H/D and S_V/D fall between 1.25 and 3.0.

The advantages of pin fins over conventional fins can be seen from the results of a controlled experiment. Two power supplies, one with conventional extruded fins and one with pin fins, were operated under the following conditions: input 115 V, 60 Hz; outputs, 5 V at 20 A, 15 V at 2 A, and 15 V at 2 A; total output power 160 W; ambient temperature 25°C.

Temperatures measured at nine locations showed that, while the tempera-ture rose an average of 52°C on the conventional fins, it rose only 40°C on the pin fins. When the load on the pin fin power supply was increased until the average temperature rise was 52°C (the same as for the conventional fins), the output became 5 V at 20 A, 15 V at 3 A, and 15 V at 3 A, for a total output of 190 W.

A 30-W increase was possible by incorporating pin fins in place of the conventional extruded heat sink. Thus, with temperature the limiting factor, the pin fins allowed the power supply's design to deliver 18.75% more power (or to operate at the same power level with significantly improved reliability).

FORCED-AIR COOLING

Although the low-velocity airflow resulting from convection cooling is adequate for many low to medium power supplies, high-power density units require additional cooling. Forced airflow applied to conventional finned heat sinks will provide significantly better cooling than natural convection airflow. For example, a convection-cooled 600-W switcher, when cooled by forced air, can thermally handle about 750 W, even with its volume reduced 30% from 710 in.^{3} to 420 in.^{3}.

Most high-wattage power supplies (500 W and above) are designed with fan cooling. Since the fan is chosen to handle the thermal demands of the supply, the system designer need only ensure that airflow between the supply and the outside of any enclosures is not restricted. If the airflow is restricted, the fan will circulate the air at reduced velocity, resulting in lower cooling efficiency.

It is not unusual, especially when using switching-type supplies, for the load to generate a significant proportion of the entire heat dissipated within a system enclosure. In some cases, the air drawn through the system by the power supply's fan is enough to cool both the supply and the load. However, when the supply is convection cooled or when its internal fan cannot cool both the supply and the load, it is often necessary to use a separate fan to cool the system enclosure and ensure that overall reliability is not lowered due to poor thermal management.

A good approximation of the cooling fan required can be obtained by making a few simple estimates of the system's power dissipation, the airflow needed, and the resulting pressure drop.

The power dissipated by a system is the sum of the load and the power lost due to the inefficiency of the power supply. It can be calculated simply by dividing the output of the power supply (the power consumed by the load) by the supply's efficiency:

$$P_{dis} = \frac{P_{out}}{\eta} \qquad (5.13)$$

The airflow needed to properly cool the system is a function of the power dissipated, the mass of air, and the amount of heat that air will hold. The thermodynamic relationship between these factors is expressed by

$$P_{dis} = MC_p \, \Delta T \qquad (5.14)$$

where M is the mass of air per unit time, C_p is the specific heat, or thermal capacity, of the air at a constant pressure, and ΔT is the temperature rise of the air.

Substituting in the values of M and C_p at sea level and simplifying yields

$$\text{cfm} = \frac{(1.76)P_{dis}}{\Delta T} \qquad (5.15)$$

where cfm is the airflow in cubic feet per minute, P_{dis} is the total power dissipated in watts, and ΔT is the temperature rise of the air in degrees Celsius.

When specifying a cooling fan, pressure drop as well as cfm must be

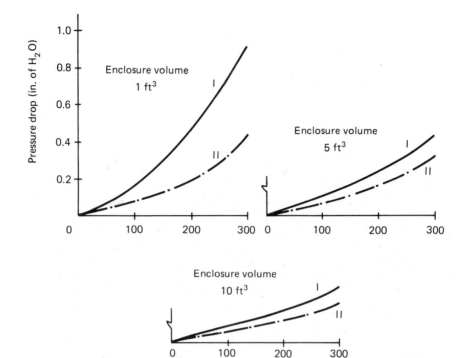

FIGURE 5.5.
Estimated pressure drops for (I) densely packaged enclosures that restrict about 75% of the airflow and (II) lightly packaged enclosures that restrict about 25% of the airflow. The air intake and exhaust vents are 16 in.^{2} for all three enclosures.

quantified. The pressure drop is a function of the size and physical layout within the enclosure. The more densely packed the enclosure, the higher the pressure drop. The exact pressure drop for an enclosure is difficult to calculate but can be estimated from empirical data (Figure 5.5).

The graph of static back pressure versus airflow rate for most enclosures is parabolic. The pressure drop is proportional to the square of the airflow velocity in cubic feet per minute. A curve can be established for a given system at all flow rates after the system resistance has been determined.

Many manufacturers of fans have calibrated units with test curves available for empirically determining the system resistance. The calibrated fan is installed in the cabinet and the airflow rate monitored. The system resistance and point of operation·can be extrapolated from the calibrated performance curve once the airflow rate is determined.

Once the system resistance curve of pressure versus airflow is plotted, the characteristic curves of several fans can be superimposed on it. The fan whose most efficient operating region intersects the airflow–back pressure curve is the preferred choice for the system.

HYBRID
THERMAL DESIGN

A relatively new development in power supply design is the hybrid thermal design. In conventional fan-cooled power supplies, the heat-producing transistors or diodes are mounted on a heat sink that is electrically *and* thermally isolated from the chassis. The hybrid approach has the heat-producing devices mounted in such a fashion that they are electrically, but *not* thermally, isolated from the chassis.

By being able to use the relatively large chassis surface for convection cooling, in addition to the continued use of a fan for internal forced-air cooling, power supplies that employ the hybrid thermal approach can achieve higher power densities and improved reliability.

Careful planning and a clean mechanical design are necessary to effectively use hybrid cooling, but the added benefits are well worth the effort. A typical (5 by 8 by 11 in.) medium power switcher has an outside surface area of 325 in.^{2}. That much surface area of 1/8-in. aluminum chassis material can dissipate heat very effectively without a significant temperature rise.

Consider an 80%-efficient 750-W switcher with hybrid cooling in a 5 by 8 by 11 in. package with the thermal load split 80%–20% between the forced-air and convection mechanisms, respectively. The power supply dissipates 187.5 W total, of which 37.5 W is through convection using the case. Given the surface area of the case as 325 in.^{2} (some surface is lost to the fan intake, terminal blocks, air exhaust vents, etc.), the heat dissipated per square inch is 0.115 W/in.^{2} (37.5 W/325 in.^{2}). Finally, using the Equation (5.12), the case temperature rise is given by $\Delta T_c = [125°C/(W/in.^2)] \times 0.115 \ W/in.^2 = 14.4°C$, a minimal temperature rise to gain 20% additional cooling. The thermal improvement is roughly equal to the relative gain derived from using pin fins over conventional extrusions for standard convection cooling.

GENERAL SYSTEM
LAYOUT

The internal layout of an enclosure cooled by natural convection must encourage the free flow of the convective cooling current. To minimize the internal temperature rise, both the power supply and load must be considered.

Make sure that all system components that dissipate significant amounts of power are not crammed into corners or blocked by adjacent components so that hot spots develop within the enclosure. Wherever possible, isolate heat-producing components from each other. The goal should be to locate these components so that a nearly uniform average temperature rise occurs over the entire enclosure's outer surface. If this procedure is followed, the entire system will run cooler and be more reliable.

When using forced-air cooling, both the placement of the fan and the direction of airflow are important in determining system thermal performance. If the fan is used to exhaust air from the enclosure, the interior will have a negative pressure. The enclosure will be sucking air in through every crack, as well as through the filtered air intakes. While the overall volume of airflow will be improved, dust and dirt will find their way into the enclosure. Using the fan to blow air into the enclosure creates a positive internal air pressure and prevents the inflow of dust and dirt. At the same time, the increased pressure causes the air inside the enclosure to be denser, providing a better cooling medium.

In most practical system enclosure layouts, the heat-generating components are located in a number of different locations (sometimes even in hard to cool corners). Because of this, the best way of cooling the system might be to use individual air ducts to carry cooling air from the fan to each component that needs it, varying the ducts cross-sectional area to serve each individual component's heat load. This is often done for power supplies in systems that use a number of high-wattage units. As cooling-efficient as this scheme may be, it is complicated, wastes valuable space, and is expensive. It is generally reserved for large, complex systems that generate large amounts of heat.

A good compromise for small and medium systems is the use of baffling and exhaust vents for heat-generating components to direct their airflow to some central air path. The positive air pressure provided by the fan forces the air to flow around the baffles and past the heat sources on its way out through the exhaust vents. An important consideration when using this approach is to control the restrictions on airflow and to place the vents to ensure adequate airflow where it is most needed.

WHAT ASSUMPTIONS?

It is important to note that this discussion of reliability considerations has been on a practical level and is not theoretically rigorous. First, the Stefan–Boltzmann law was linearized, then all radiating surfaces were assumed to have an emissivity of 0.9, system operation was assumed to take place at sea level, temperature gradients do exist, and so on.

The linearization of the Stefan–Boltzmann law will generally give results accurate to within 15% over the range of 0 to 0.4 $W/in.^2$, which is close enough for most applications. The emissivities of most materials used for electronic

system enclosures and chassis are surprisingly close to 0.9; the pin-fin chassis discussed on page 65 has a measured emissivity of 0.87.

For most applications, the assumption about sea-level operation will be sufficient. Finally, the assumption of uniform average temperatures is not too excessive, provided that the system enclosure is properly laid out to minimize them so the computed average has some meaning.

SUMMARY

Some power supplies are touted by their manufacturer to deliver "more watts per cubic inch." It may prove to be a hollow claim, however, since a high power density supply may also produce the most heat. Consider a 30%-efficient linear with a rated output of 10 W packaged with a radiating surface of 50 in.^{2}. While this corresponds to a modest 0.2 W/in.^{2} unit power output, the case will experience an average temperature increase of 60°C. Low efficiency and/or a small radiating surface have made this supply a thermal disaster that is a prime suspect for early catastrophic failure.

By the same token, high power density, along with a low case temperature rise, are indicative of good thermal design and a reliable power supply. Usually, an efficient power supply with good thermal design can operate more reliably at higher temperatures without derating. The bottom line is not whether switchers are more or less reliable than linears based on predicted MTBFs, but rather whether any particular power supply is reliable enough to power the system load circuits through their useful lives.

Even though an MTBF calculated using the component failure rate data in MIL-HDBK 217 and taking thermal parameters into consideration

TABLE 5.3

Comparison of Load and Power Supply
MTBFs Using 5-V, 100-A Power Supplies
per MIL-HDBK 217

Device Family	λ	N	$N\lambda$	MTBF (hours)
TTL	1.05×10^{-6}	6.67×10^9	7,000	140
ECL	1.50×10^{-6}	3.33×10^9	5,000	200
Memory	3.90×10^{-5}	2.00×10^9	78,000	13
Linear	—	—	—	28,000
Switcher	—	—	—	18,000

cannot be applied with certainty to a specific unit, it is still a statistical bench mark for a population, whether it consists of loads or power supplies. Using the data in MIL-HDBK 217, it is possible to calculate the MTBF for the load as well as the power supply; if the MTBF of the power supply is greater than that of the load, the load can be expected to fail before the power supply does. Since reliability costs money, it is generally wise not to buy any more than is needed by any particular application.

Although actual digital loads consist of a mixture of IC families, a theoretical load can be constructed for each technology by determining the average current drawn by an IC and dividing that into the maximum rated output current of a power supply.

Table 5.3 compares the calculated MTBFs of TTL, ECL, and memory ICs with those of typical 5-V, 100-A switcher and linear power supplies. The results of this comparison show that most power supplies can be expected to outperform their loads by a comfortable margin.

QUALITY

A reliable design is necessary, but not sufficient, to ensure a quality power supply. For the
calculated MTBF of any power supply design to
become a reality, the supply must be properly
built. MTBF calculations are simply that, cal-
culations. They do not take into consideration how components are selected,
what the factory's QA and QC capabilities are, or how the unit is burned in and
tested. These areas, taken together, determine the quality of a power supply. A
high-quality power supply is one for which the actual MTBF has a high proba-
bility of matching or exceeding the calculated MTBF. A *low-quality* power
supply will likely exhibit an actual MTBF much lower than the a priori
calculations.

Power supply burn in and testing are critical areas determining the quality
level of a given unit. If the burn-in period is not long enough or stressful enough,
all component deficiencies may not be experienced, and early random failures
can be excessive. Without accurate and complete testing, marginal units may
find their way into systems and result in poor performance or nonperformance of
the final product.

73

As the complexity of power supplies continues to increase, quality is becoming a more elusive parameter to determine. Linears, with relatively simple circuitry, can be subjected to a correspondingly simple testing and burn-in cycle to achieve a given level of quality. Switchers, on the other hand, are subject to a wider variety of complex failure modes and must be subjected to closer scrutiny to yield an equivalent level of quality.

COMPONENT-LEVEL CONSIDERATIONS

The amount of care given to component selection and screening says a great deal about the expected quality of the finished power supply. For example, MIL-HDBK 217 calculates stress on power semiconductors by comparing the worst-case operating junction temperature to the maximum rated junction temperature. Second breakdown, a primary cause of power transistor failure, is not taken into consideration.

Second-breakdown failures usually occur well within the safe operating power specification of the transistor. They are irreversible and almost instantaneous. Their prevention depends almost completely on knowing the precise limitations of the specific transistor and thoroughly analyzing all its operating conditions. Paying close attention to transient and short-circuit situations, the power supply designer must know variations of semiconductor manufacturer's specifications for second breakdown and must apply those specifications correctly. New devices, even from known suppliers, must be thoroughly qualified by designers prior to usage, since an error in application can substantially reduce the realized MTBF of a design.

A second cause of failure in power transistors, especially in linear supplies, stems from the unequal thermal expansions of the chip itself and the spreader that connects it to the header of the transistor case. High-current devices are more prone to this failure mode than low- to medium-current devices. The relatively large semiconductor chips required for high-current (over 25 A) devices increase the potential for this failure mechanism.

Thermal stresses can be substantially reduced by placing a molybdenum pad between the semiconductor chip and the spreader. This type of construction tends to increase the transistor's price. As a result, these transistors are not used in low-cost power supplies. If the power supply manufacturer chooses not to pay the monetary price for the better device, the user pays a price in lower reliability. Internal construction, as well as electrical characteristics, play an important part in overall transistor reliability.

MIL-HDBK 217 calculates capacitor stress by comparing the maximum dc working voltage (wV dc) and temperature ratings to the actual circuit parameters. In many power supply designs, the most severe stress comes from the *ripple current*, not the working voltage or temperature. Ripple current through an electrolytic capacitor causes heating of the electrolyte. If the electrolyte heats sufficiently and vaporizes, the capacitor is destroyed. When operating at temperatures below the vaporizing temperature of the electrolyte, useful capacitor life is affected by equivalent series resistance.

The *equivalent series resistance* (ESR) further complicates the process of translating calculated power supply MTBFs into reality, since ESR degradation for a given temperature varies from one grade of capacitor to another, even within a particular manufacturer's line. Computer-grade capacitors generally have a lower ESR than TV-grade components, and 85°C computer-grade components are superior to 65°C components. Most power supply manufacturers do not closely monitor the ESR of the electrolytics they purchase, relying rather on the assumption of lot-to-lot consistency from the capacitor manufacturer.

To see the variations possible for the "same" component from various manufacturers, gather up a selection of power transistors with the same JEDEC number and a collection of identically rated electrolytics. Cut the top off the transistor cases and compare them for wire gage, thoroughness and cleanliness of lead attachment to the output terminals, and overall uniformity. When examining the electrolytics, look at the end seals, packaging density, CV product, ripple-current capability, general construction, and, of course, ESR. The number of variations found in the "same" product from different manufacturers can be amazing.

Even the lowly resistor is not as simple as it might seem at first. Variable resistors have a critical place in the overall performance of power supplies. Any change or instability in the voltage adjustment or other potentiometers will be reflected on the output. Resolution, torque, and stability are all important to reliable power supply performance, and none are taken into consideration when calculating MTBF according to the guidelines of MIL-HDBK 217. Did the manufacturer use inexpensive (and less desirable) single-turn, open-construction potentiometers or the more expensive and reliable multiturn, sealed variety?

Component screening is also a key factor. It is important that all critical components, such as power semiconductors and capacitors, be closely looked at by the manufacturer's incoming inspection. Some manufacturers burn in all semiconductors and then 100%-inspect the survivors to screen out potentially weak links; others do no burn in and only inspect a sample at incoming inspection. The effect on field reliability can be significant.

Not only is component screening important; the parameters tested are also an important factor. It would be absurd to 100%-inspect power transistors or filter capacitors if the only test consists of counting the leads and rejecting those parts that do not conform. The critical parameters such as ESR in capacitors or characteristic curves and switching speeds in semiconductors must be examined if the inspection is to have any significant result in improved reliability and higher quality.

At this point the design has been checked to determine its calculated MTBF, and the components have been examined to ensure that they are of sufficient quality. The final step is to ensure that the execution of the design is reasonable. The electrical and mechanical manufacturing techniques must now be monitored to ensure a quality product. This is where the process of product burn in and testing takes over.

BURN IN

The primary purpose of the burn in of power supplies, or any electronic system, is to force the occurrence of all early failures under controlled conditions prior to sending the unit into the field. A thorough and complete burn in will just reach the bottom of the "bath-tub" curve, but will not significantly extend into the bottom portion of the curve. The term *burn in* indicates that a unit is operated at an elevated temperature for a period of time. It is therefore common for power supply specifications to define the burn-in period in terms such as "operation at full load and 50°C for 8 hours."

Power supplies, however, are subject to a variety of stresses not simulated in the typical burn-in test. By specifying burn in in the standard and relatively simple way, not all the early failures are consistently weeded out. *Stress-aging* is a better term for the process, since it can include all important stresses needed to properly "burn in" a power supply. In fact, the type of stress applied is more important than the duration in determining the number of units that will fail during burn in (Figure 6.1).

Most units fail upon initial turn on, when they see the first peak load or voltage transient on the input line, when they first reach operating temperature, and so on. Successive applications of a particular stress result in fewer and fewer failures, with the vast majority of failures occurring during the first few stress cycles.

The particular stresses that are important in power supply failures include initial turn on, peak loading, temperature, ac input extremes and transients, vibration, and load cycling. The implementation of each stress-aging test will depend, to some extent, on the environment in which the system will be expected to survive.

AC input cycling can vary from 30 min on/30 min off to 15 min on/5 min off. It should have an on period long enough to allow the power supply to come up to normal operating temperature and an off period long enough for it to return to ambient. This test will test the input section of the power supply. It should be included in most stress-aging programs for 100% of the production units.

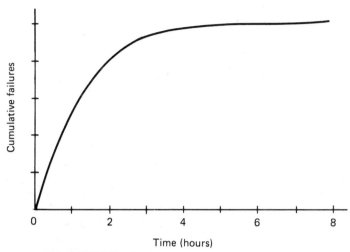

FIGURE 6.1.
For any given stress-aging test, most
failures occur when the stress is first applied. Successive application of the same
stress over time produces relatively few additional failures.

Output load cycling should be at least as stressful as expected in the system
environment. The more often and extreme the cycling, the more rigorous the
test. The regulation and output components are stressed by output load cycling.
It is applied as required by the specific system environment.

High-temperature operation, the classical burn-in test, can be taken one step
further on units with built-in thermal protection. Rather than only operating the
unit at maximum rated temperature, run the unit at an even greater temperature
to check the thermal protection circuit and make certain it can properly recover
upon cool down. Units without thermal protection should be operated at their
maximum full-load rating maximum temperature and high and low input voltage
ratings. All components are stressed by this type of temperature test. High-
temperature operation should be included in all stress-aging programs.

High-temperature turn on is used during the stress-aging process after the
power supply has achieved thermal equilibrium; the ac input power is cycled
rapidly while the unit is under full load. The ability of the power supply to
operate through dropouts in the ac line and other rapid on–off cycles is checked
with this test, which stresses the input section and should routinely be included
in stress-aging programs.

Shock and vibration testing can be as simple as a solid rap with a mallet to shake things loose or a session on a more complex three-axis vibration table. Even units that will operate in stationary systems should be subjected to a vibration test during qualification testing to ensure the mechanical integrity of the design. If they are not subjected to one prior to shipment, the carrier will most probably subject them to one during shipment.

The complete stress-aging tests discussed here are not generally employed by power supply manufacturers or original equipment manufacturers (OEMs). More often than not, the only stress applied is high-temperature operation. The OEM is given the opportunity to perform the other stress-aging in his system out in the field.

The costs of field failures and the particular systems operating environment should be analyzed to determine which stress-aging tests are to be performed. As might be expected, rigorous stress-aging of power supplies will increase the unit cost. If approached in a well thought out manner, the costs of field failures can be balanced against the costs of added stress-aging and a reasonable compromise established.

GENERAL TESTING PROCEDURES

The following discussion outlines the general procedures for testing all major power supply specification parameters. The specifications preceded by an asterisk are those normally included in incoming inspection test programs and are discussed in more detail under the heading *Subtleties of Power Supply Testing.*

The ac input source should have a low impedance. The autotransformer should have a current rating at least 200% of the actual rms input current (to help prevent clipping or distortion of the input wave form when testing switchers.)

Power supplies generally have very low output impedances (<1 mΩ), and the terminations for power and sensing should be made with care. The use of clip leads or similar terminations should be avoided. The measurement instruments should be connected to the sense leads with twisted pairs or shielded cable to avoid coupling and pickup problems.

The general test conditions, unless stated otherwise, are the following:

1. Temperature: 25°C ambient.

2. AC input: nominal, measured at the power supply terminals.

3. Output loading: full rated output current.

4. Test equipment: accuracy and stability at least 10 times greater than required resolution of test. Allow to stabilize before making measurements.

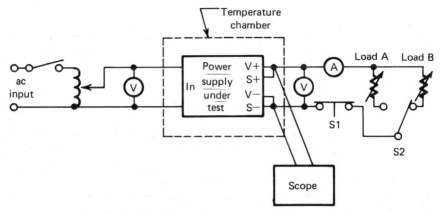

FIGURE 6.2.
General test instrumentation setup.

Warm-up drift (Figure 6.2)

1. Apply nominal ac input and measure the output voltage, V, 15 s after turn on (T_1) and after the specified warm-up period (T_2), usually 15 to 30 min.

2. $\text{Drift} = \dfrac{V_{T1} - V_{T2}}{V_{T2}} \times 100\%$

*Efficiency (Figure 6.2)

1. Measure the input current and voltage and the output current and voltage.

2. $\text{Efficiency} = \dfrac{(I_{out})(V_{out})}{(I_{in})(V_{in})} \times 100\%$

Output voltage tolerance (Figure 6.2)

1. Measure the output voltage, V_{out}.

2. $\text{Tolerance} = \dfrac{(V_{out} - V_{nom})}{V_{nom}} \times 100\%$

*Line regulation (Figure 6.2)

1. Measure the output voltage, V_{out}.

2. Vary the ac input over the specified input range and record the maximum *change* ΔV_{max} in V_{out}.

3. Line regulation $= \dfrac{V_{nom} - \Delta V_{max}}{V_{nom}} \times 100\%$

*Load regulation (Figure 6.2)

1. Adjust the load for full rated output current and measure the output voltage, V_{fl}.

2. Adjust the load for minimum rated output current and measure the output voltage, V_{ml}.

3. Load regulation $= \dfrac{V_{fl} - V_{ml}}{V_{fl}} \times 100\%$

*PARD: noise and ripple (Figure 6.2)

1. To measure rms values, connect an ac-coupled true rms voltmeter to the output.

2. To measure peak to peak, connect an ac-coupled oscilloscope to the output.

Transient response time (Figure 6.2)

1. Set load A to 25% and load B to 75% of full rated load.

2. Connect a dc-reading oscilloscope across the power supply output terminals. Trigger the scope externally on the switching of S2.

3. Switch S2 between loads A and B at the rate specified for the power supply under test, typically, 120 Hz or 5 A/μs.

4. Measure the time (μs or ms) for the scope trace to return and stay within the tolerance imposed by load regulation.

Temperature coefficient (Figure 6.2)

1. Measure output voltage, V_{nom}, at normal ambient temperature, T_{nom}.

2. Set temperature to lower operating limit, T_{low}, allow supply to stabilize for 30 min, and measure output voltage, V_{low}.

3. Set temperature to highest specified operating limit, T_{hi}, allow supply to stabilize for 30 min, and measure output voltage, V_{hi}.

4. $\text{Tempco} = \dfrac{V_{\text{nom}} - V_{\text{low}}}{V_{\text{nom}}(T_{\text{nom}} - T_{\text{low}})} \times 100\%$

or

$\text{Tempco} = \dfrac{V_{\text{nom}} - V_{\text{hi}}}{V_{\text{nom}}(T_{\text{hi}} - T_{\text{nom}})} \times 100\%$

whichever is highest.

Output impedance (Figure 6.3)

1. Measure the superimposed voltage, V_s.

2. Compute the ac component of I_L, I_L^{ac}.

3. Output impedance: $z_o = \dfrac{E_o}{I_L^{ac}}$

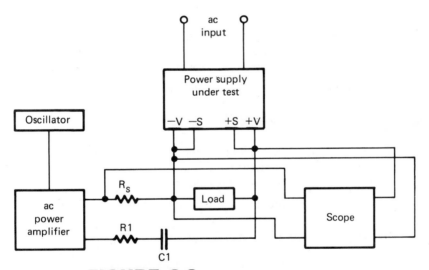

FIGURE 6.3.

Output impedance test setup. Set load at one-half full rated load. R1 should match the impedance of the ac power amplifier. C1 should yield a time constant of 2 ms in combination with R1 and should be able to carry the ac load current, $I_L{}^{ac}$. R_S can be twenty-seven 2.7-Ω, 1-W resistors in parallel to minimize inductive loading effects at high frequencies.

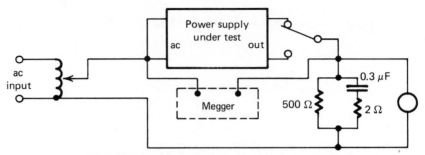

FIGURE 6.4.
Leakage current and breakdown voltage
test setup. For the breakdown voltage test,
the ac input line, output *RC* network, and
voltage meter are not needed.

Leakage current (Figure 6.4)

1. Measure the ac voltage, V_e, across the 500-Ω resistor.
2. Leakage current, $I_e = 2V_e$, in milliamperes.

Breakdown voltage (Figure 6.4)

1. Sequentially connect a high potential breakdown tester (hi-pot) or megohmmeter between the input and each output, gradually turning up the voltage until rated breakdown voltage is reached.
2. The supply fails the test if any combination of input to output fails.

Remote sense (Figure 6.2)

1. Adjust the cable voltage drop for the maximum amount that the supply is specified to compensate for, typically 0.25 to 0.50 V.
2. Measure the voltage at the load, V_e.
3. The load voltage, V_e, should be within the output voltage tolerance limits specified for the supply.

Remote inhibit (Figure 6.2)

1. Measure the nominal output voltage to ensure proper functioning of the supply.

boost is dependent upon the specific test setup: the autotransformer, how it is wired, the peak currents involved, and the wiring impedances.

The nonsinusiodal input voltage and current wave forms encountered when testing power supplies can make measuring power supply efficiency quite difficult. The output power, being dc, can be measured with extreme accuracy, but since the input is nonsinusoidal, the true integration of $(1/T)\ di/dt$ to determine the input power can be very complex. Short of purchasing some very expensive instrumentation, at least three ways can be used to measure the input power and efficiency of a power supply to within $\pm 5\%$.

For power supplies that do not have input transformers, such as off-line switchers, dc power can be applied to the input terminals and the input power directly measured. The error factors involved in this method derive mostly from the fact that the input rectifiers are seeing dc instead of pulsed ac, and there is no ac loss in the input capacitors.

Graphical analysis of the input ac wave form can be used on any power supply to yield reasonably accurate estimates of input power. Being careful with regard to sweep starting points, an oscilliscope can be used to picture both the voltage and current input wave forms. Using about 20 to 25 points over one half-cycle of input line frequency to perform a graphical integration of voltage and current can yield some very accurate estimates of ac input power.

A less direct method is to use calorimetric techniques to measure the amount of the power dissipated and then add that to the output power to determine the total input power. A fairly accurate calorimeter can be constructed with easily obtained materials, such as a Styrofoam ice chest, Freon, and so on.

The specification of current limiting can be an area of confusion between power supply manufacturers and users. A typical specification might read, "The current limit set point is 115% to 120% of full load. Current limit is the foldback type with short-circuit current limited to a maximum of 40% of full load." This sounds straightforward enough, but what is the "set point"?

Most users test current limit by measuring the peak current and verifying that it is less than some maximum value. Unfortunately, that maximum current is not what is referred to as the "set point." In fact, peak current is not generally specified at all. The "set point" is defined differently by each manufacturer. It is used as an indication that the current-limiting circuitry has been activated, not that peak current has been reached.

Once the current-limiting circuitry is in operation, the peak current will follow. However, in power supply manufacturing it is easier to control the set point than it is the peak current level. Commercial component and manufacturing process tolerance buildups cause peak current to be difficult to accurately predict. The *set point* is generally defined as a certain drop in output voltage as the load current is increased beyond the rated maximum. For example, the set

2. Implement the remote inhibit function (usually by connecting the remote inhibit terminal to the negative sense terminal.)

3. Measure, $V_{out} = 0$.

*Power-fail detection for switchers only (Figure 6.2)

1. Connect one trace of a dual-trace dc-reading oscilloscope across the power-fail detection output. Trigger the scope externally on the turn off of ac power.

2. Connect the second trace across the power supply output terminals to monitor the output voltage.

3. Measure the time for the power-fail signal to be triggered relative to the output voltage dropping out of the regulation band (Figure 6.5).

*Holdup time for switchers only (Figure 6.2)

1. Connect a dc-reading oscilloscope across the power output terminals. Trigger the scope externally on the turn off of ac power.

2. Measure the time (ms) for which the output remains within the regulation band after removal of ac power. *Note:* This parameter may be specified at low line or nominal line; check the specification for the particular power supply.

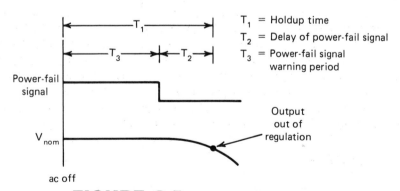

FIGURE 6.5.
The timing of the power-fail signal is measured relative to the loss of ac input and the loss of output regulation.

*Overvoltage protection (Figure 6.2)

1. Gradually increase the external voltage source, V_{ex}, until the supply's OVP fires.
2. Measure V_{out} at the OVP point.

*Current limit point (Figure 6.2)

1. Gradually increase the load until the current limit circuit in the power supply is activated.
2. Measure current at the current limit point, I_{lim}.

*Short-circuit current (Figure 6.2)

1. Connect current meter directly across power supply output terminals.
2. Measure short-circuit current, I_{sc}.

Load interaction or cross regulation (Figure 6.2)

1. Measure V_{out} for each output at each step of the test pattern.
2. Load interaction or cross regulation is the farthest excursion from nominal, expressed as a percentage:

$$CR = \frac{V_{nom} - \Delta V_{max}}{V_{nom}} \times 100\%$$

Stability (Figure 6.2)

1. Apply nominal ac input and measure V_{out}, the output voltage, after the specified warm-up period (T_1), usually 15 to 30 min.
2. Hold the supply at a constant temperature and remeasure V_{out} after 8 hours (T_2).
3. Stability $= \dfrac{V_{T1} - V_{T2}}{V_{T2}} \times 100\%$

SUBTLETIES OF POWER SUPPLY TESTING

Special considerations can come into play when testing power supplies, especially off-line switchers. Areas that must be considered include ac input volt-

ages and wave forms, efficiency, current limiting, short-circuit current, an ripple and noise.

Switchers do not draw line current until the input voltage exceeds the holding voltage across the input capacitors (Figure 6.6). As a result, current flows in the input line for only about 2 ms for each half-cycle of ac. This causes a unit with a 16-A average current draw to have a peak current draw of about 67 A. In the real world, when the power supply is in the final system, drawing input current from an electrical outlet or other low-impedance source, the high peak current does not cause an input voltage problem.

When testing power supplies, however, the input current is typically drawn from an autotransformer, not directly from the ac line. The use of the autotransformer allows the input voltage to be varied over the specification limits to ensure thorough testing of the unit, but it also introduces a possible source of error when measuring input voltages. The high peak current draw of a switcher, together with the relatively high impedance of the autotransformer, causes the voltage form factor to change and will result in an incorrect voltage reading if an rms meter is used.

This change in the relationship between rms and peak voltage measurements can easily be 5% or greater. In the case of a close design specification, this is enough to cause failures in line regulation or voltage sequencing tests. To avoid this problem, it is necessary to either use a peak voltage meter or boost the acceptable rms value enough to compensate for the input effect. The amount of

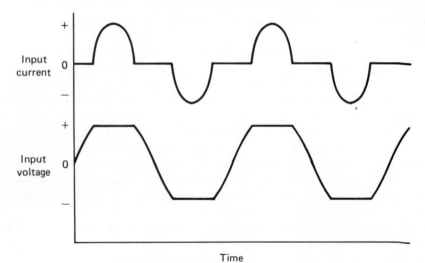

FIGURE 6.6.
Input current in off-line switchers flows only during the period of peak input voltage and tends to clip the voltage peaks.

point may be defined as the output current level at which output voltage is down 0.6% from the no-load voltage (Figure 6.7).

When testing the current-limiting circuitry, it is important to have the manufacturer's specific definition of the current limit set point. If he measures voltage roll-off instead of peak current, you can ask what the worst-case peak current is expected to be. With that information, you can proceed to test for either voltage roll-off or (with somewhat less accuracy) current peaking.

The activation of the current-limiting circuitry is only one-half the story if foldback current limiting is used in a particular power supply. The second consideration is the short-circuit current produced by the supply. The problems encountered here are easier to resolve than was the case with the current limit set point. A typical power supply specification does not adequately define the short-circuit current; both upper and lower bounds are needed for defining acceptance criteria. Generally, only the maximum short-circuit current is stated.

Once the upper and lower limits are set, the method for performing the test must be examined. The required test condition is 0 V (an ideal short) at the output of the supply. Depending on the specific test setup and current levels involved, this requirement may be violated enough to produce inaccurate test results. Consider the following example:

$$\text{Full load current, } I_{fl} = 300 \text{ A}$$
$$\text{Peak current, } I_p = 345 \text{ A}$$
$$\text{Maximum short-circuit current, } I_{scm} = 100 \text{ A}$$
$$\text{Voltage at peak current, } V_1 = 4.9 \text{ V}$$
$$\text{Voltage under short circuit, } V_2 = 0.25 \text{ V}$$

Assuming linear operation of the current-limiting circuit in foldback mode from I_p to I_{scm}, we have

$$I_{scv} = I_{scm} - \left(\frac{I_p - I_{scm}}{V_1 - V_2}\right) V_1$$

In this case, we have

$$I_{scv} = 100 - \left(\frac{345 - 100}{4.9 - 0.25}\right) 0.25 = 87\%$$

In this example, the testing techniques used introduced such a serious error factor that any units that are greater than 87% of the specified maximum short-circuit current will be rejected, a 13% error. It is also true that units 13% below the minimum short-circuit current will be accepted.

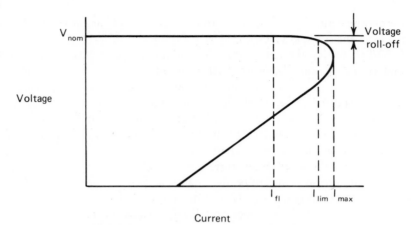

FIGURE 6.7.
When measuring current limiting, three current levels are important: full load current, I_{fl}, set point (activation) current, I_{lim}, and maximum current, I_{max}.

This effect is greater the higher the peak current and the higher the departure from a true 0-V short circuit. For power supplies under about 500 W, these effects can often be ignored. For higher-power units, such as the 1500-W unit used in the earlier example, the effect of less than ideal test conditions can be very serious and must be dealt with. The voltage drop departure from zero should be minimized by placing the short circuit *directly* across the output terminals whenever possible. The maximum terminal voltage should be less than some acceptable value, such as 0.010 V.

A final area of difficulty, particularly when testing switching power supplies, is the specification of PARD. The term PARD, an acronym for "periodic and random deviations," was coined because the differentiation between "noise" and "ripple" in a power supply is not necessarily obvious or meaningful. If noise is random and ripple periodic, the high-frequency spikes on the output of a switcher must be ripple. Most engineers, however, refer to them as noise.

To eliminate this source of confusion, the term PARD is used to refer to the peak-to-peak or rms value of *all* ac components present on the dc output voltage. PARD is typically defined either as a given voltage (e.g., 50 mV) or as a percentage of the nominal output voltage (e.g., 1%). Although no bandwidth is generally specified, it is assumed to be dc to 30 MHz unless otherwise stated.

The main components of switcher PARD are the *line effect* (strongest around 120 Hz), the *switching frequency effect* (strongest at 20 to 50 kHz), and the *switching transient effect* (strong from 1 to 20 MHz). The line and switching effects are sinusoidal, with switching effect superimposed on the line effect, while the switching transient effect appears as a series of spikes on top of everything else.

When testing PARD, it is important to remember that radio frequencies are involved and spurious pickup is a constant possibility. To minimize these problems, PARD must be measured directly at the output terminals using a 1 : 1 probe and the shortest possible lead length outside the grounded cable. It can be instructive to have the PARD of a single switcher measured by different people using different test gear. The variations in test results can be surprising.

POWER SUPPLIES AND AUTOMATED TEST EQUIPMENT

The transition from linears to switchers is having some interesting side effects. With switchers replacing linears in virtually all electronic systems, computerized or automated testing of power supplies is becoming more and more commonplace. As the demands on power supplies grow, so do the demands on testing.

Even a catalog standard switcher can be required to sequentially power up its multiple outputs, maintain the output voltage for a specified time after loss of input voltage, be able to change its output voltages as directed by an externally applied signal, or provide a failure signal when the line voltage is lost, as well as an OK signal when it is restored. Traditional, manual test-bench setups are often hard pressed to efficiently and accurately test the growing number of complex parameters associated with power supplies.

To further complicate matters, a comprehensive test program must include line and load regulation, ripple and noise, overvoltage protection, current limiting, and short-circuit current, as well as any other measurements peculiar to a specific application. As if that were not enough, most companies, whether power supply manufacturers or OEM buyers, deal with a number of different power supplies, each with its own set of specifications, test requirements, and (unfortunately) design changes.

Automation is becoming a driving force behind many industries. The same is true, for better or worse, in the power supply industry. Automated testing by power supply manufacturers is both a blessing and a curse for power supply users. That automated test equipment (ATE) lowers costs (and one hopes prices) and improves the accuracy and consistency of testing are certainly benefits to the purchaser. However, the lack of standardization of ATE in the power supply industry can cause the original equipment manufacturer (OEM) significant problems.

Most problems associated with ATE for power supplies relate to the dynamic nature of power supply testing. When considering load or line regulation, PARD, efficiency, or other similar "static" parameters, there are usually no problems with automated testing. However, when considering dynamic parameters such as output sequencing, holdup storage, power-fail signal timing, input inrush current, or transient load response, the comparability of various testers can leave much to be desired.

Holdup time is one area in which correlation problems can arise between automated and manual testing. All ATE systems use a triac or similar device to switch off the ac input when measuring holdup time. These electronic switches operate at the zero crossover point and yield repeatable measurements. In a manual test setup, the ac input is interrupted with a mechanical switch. The difference between an asynchronous mechanical switch and a zero crossover electronic switch can cause significant correlation problems.

The measurements of holdup time using the zero corssover method (ATE) are averages since they take place exactly halfway between the peak input current flows (Figure 6.8). Using a mechanical switch could interrupt the ac just prior to the beginning of current flow, rather than halfway between. The error introduced by always cutting ac input at the halfway point is

$$\frac{16.67 \text{ ms} - 4 \text{ ms}}{4} = 3.17 \text{ ms}$$

This error factor is independent of the actual holdup time of the power supply. It is a result of the test equipment characteristics. When testing holdup time with an ATE system, the 3.17-ms error factor must be added to the specified holdup time. A supply with a 16-ms holdup time specification should be tested at 19.17 ms. If this correction is not made, there is a potential error of almost 20% (3.17/16.0) when measuring holdup time.

Another potential problem area is the measurement of power-fail signal timing. The timing of this signal is usually fairly sensitive to operating temperature, output loading, and ac input voltage. Consider the same quad output switcher tested on a manual and then on an ATE test setup. In the first instance (manual), the power-fail signal timing is one of the final tests in the test sequence, and the unit has been operating over a half-hour when it is performed; the unit will have reached thermal equilibrium. In the ATE system, all tests are complete within 15 min; the unit will not reach thermal equilibrium.

Output loading must be the same in both tests. If one test is performed with more loading on the auxiliary outputs, it could show a different signal timing. The loading is important since the auxiliaries in multiple-output switchers often include linear postregulators and are, therefore, less efficient than the primary output. An ampere of load on the auxiliaries will draw more energy from the input capacitors than an ampere of load on the primary. As a result, the holdup time and power-fail signal timing will be directly affected. This is not so much a problem of manual versus ATE as it is a problem of communicating *all* test parameters.

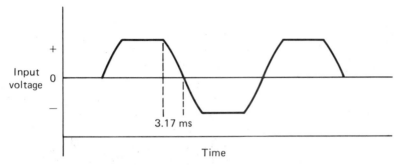

FIGURE 6.8.

The potential error when comparing holdup time measured on a manual test bench with that measured on an ATE system is 3.17 ms. The ATE system will consistently estimate holdup time at 3.17 ms less than it actually is.

The ac input voltage is important since it directly affects the energy stored in the input capacitors. The problem of ac line clipping was discussed in the previous section. It can have a direct and significant impact on the measured timing of the power-fail signal. The variations that can occur in this area are not limited to manual versus ATE, but generally exist from one test setup to the next.

The use of ATE in power supply production should definitely be beneficial for OEM users in the long term. It will mean lower cost and more consistent and reliable testing of increasingly complex power supplies, especially switchers. In the short run, however, power supply manufacturers and OEM users must work very closely together to ensure that the potential benefits of ATE can be realized.

SUMMARY

Obtaining a high-quality power supply is a complex process. Beginning with a reliable design is essential, but the process goes far beyond simple MTBF calculations. Components must be properly selected and screened, completed power supplies must be thoroughly stress-aged, and test results must be correlated between the power supply manufacturer and system manufacturer.

A key aspect of the process is adequately defining the required level of quality in terms of specific measurable parameters. These parameters generally take the form of initial design qualification and approval tests and ongoing performance testing. The particular tests performed depend on the system requirements, as well as on the type of power supply involved.

On a component level, it is important to identify how rigorously each component is tested and burned in. Critical components, such as switching or pass transistors, filter capacitors, and magnetics, must be given constant scrutiny. The specific tests performed are just as important as the choice of components to be screened.

Stress-aging is an area that is often overlooked or oversimplified when testing power supplies. When specifying a stress-aging procedure, the operating environment of the final system must be kept in mind so that it may be adequately modeled by the procedure. Which stresses are applied are more important than the duration of the process.

With both power supply and system complexity continuing to increase, the burn-in and testing process is becoming more and more difficult to efficiently implement. As a result, ATE is being used more often with power supplies, both in production and at incoming inspection. While ATE helps to increase efficiency, it demands that all test parameters be thoroughly defined to ensure that its benefits may be realized.

PROTECTION AND DISTRIBUTION

7

Increasingly common are power supplies that provide much more than simple voltage regulation. In today's complex electronic systems, power supplies are often called on to provide power-fail alarms, remote sensing and margining capability, provision for parallel operation, and a host of other interface and control signals in addition to the standard overvoltage protection and current-limiting circuitry.

Adding various support and protection functions to a power supply can enhance system capability and performance, but it also has the potential for causing more problems than it solves. Examples include overvoltage protection that does not protect sensitive circuitry, remote sensing that does not improve at-the-load regulation, and power-fail signals that send false warnings. It is wise to investigate all potential problem areas before finalizing the power supply specifications. Fully defining the important characteristics of all protection and distribution functions will prevent unpleasant surprises when the system is turned on.

PROTECTION

Overvoltage protection

Overvoltage is the most common, and often misunderstood, protection specification. Overvoltage protection (OVP) is not used to protect the power supply from surges in the ac line; it is used to protect the load circuitry from damage caused by excessive output voltage from the power supply. There are two questions to be answered when specifying OVP: Is it really needed, and what form should it take?

Whether the load really needs (or justifies) OVP depends, in part, on the maximum voltage it will tolerate and the maximum voltage the power supply is expected to deliver under a fault condition. In the case of TTL circuitry, OVP can be essential. CMOS, however, will not generally need it. Voltages of 20 V or more can be tolerated by CMOS components without damage. Even if the regulator should fail and feed high voltage to the output terminals, it is unlikely that a 5-V supply would ever produce anything approaching 20 V.

A second determining factor is the value of load circuitry. It does not make sense to spend $10 for OVP to protect $2 worth of components. The value of the load components is related to the output current of the power supply. Low-current outputs will drive relatively fewer components than higher-current outputs, and, therefore, low-current outputs are less likely to justify OVP. For output current levels of under 5 A, OVP is not usually economically justified; from 5 to 10 A the picture is unclear and depends on the particular load; above 10 A it is often easy to economically justify the addition of OVP.

Once it is determined that OVP is needed on both a technical and economic basis, it is not enough to simply request that the power supply have OVP. An often overlooked aspect of OVP, as well as other protective circuits, is the speed of activation. Many power supply specifications define OVP only in terms of a window of voltages in which the circuit is activated. There are two primary techniques used to implement OVP. Which one is used has important implications on the reaction time as seen by the load circuitry.

Linears generally incorporate what is called a *crowbar* type of OVP (Figure 7.1). In this OVP scheme, an SCR is placed directly across the output terminals. When an overvoltage condition occurs, the SCR is fired, which shorts the output almost instantly. Typical reaction times of crowbar OVP circuits are measured in microseconds. It is important that the SCR control circuit have the correct level of sensitivity. If the SCR is not fired quickly enough, a sudden rise in output voltage could harm sensitive load circuitry. On the other hand, if the crowbar reacts too fast, there can be excessive nuisance turn ons of the SCR caused by RF noise or other nonlethal transients.

Inverter-inhibiting OVP circuitry is often used in switching supplies (Figure 7.2). This type of OVP senses the output voltage and fires when an overvoltage

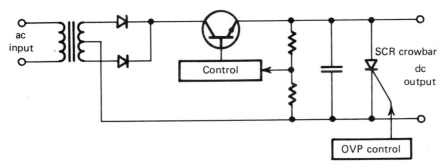

FIGURE 7.1.
When activated, the SCR is a crowbar-type overvoltage protection scheme acts as a direct short circuit across the output of the power supply.

condition occurs, just as in the case of the crowbar approach. With the inverter-inhibiting approach, the firing of the OVP does not immediately shut down the output voltage. Depending on the relative loading of the supply, the delay can be a few milliseconds or longer. When the OVP is fired, the energy stored in the output filter section must drain off into the load before the voltage goes away, and hence the delay in the OVP reaction time as perceived by the load circuitry.

In most instances, the lag in reaction associated with the inverter-inhibit approach is not a problem. When it is, the crowbar technique should be specified to ensure that any overvoltage condition is clamped off as quickly as possible. In medium- to high-wattage supplies, the inverter-inhibit OVP is generally less expensive than the SCR approach.

One last point to be aware of is that, whenever the SCR technique is specified in a switcher, an inverter-inhibit OVP circuit can be included as a

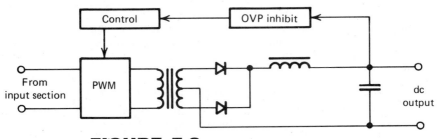

FIGURE 7.2.
Inverter-inhibiting overvoltage protection acts through the feedback-control loop to shut down the power supply in the event of an overvoltage condition at the output.

backup. This dual-OVP approach is often used. In addition to the obvious reliability benefits derived from having redundant OVP circuits, shutting down the inverter during an overvoltage condition reduces stress on the SCR, as well as the rest of the power supply, since the supply is not pushed into a current-limit mode, as would be the case with only the crowbar short circuit across the output.

Because OVP is used to protect the load from faults in the supply, it is a latching-type protective circuit. Whenever an OVP circuit fires, there is assumed to be a fault in the power supply. Therefore, the OVP circuit latches the supply off until the power is cycled off, the fault is presumably corrected, and then the power is cycled back on. To reset an activated OVP circuit, it is necessary to cycle the input power off and then on again.

Current limiting

Current limiting is another very commonly encountered power supply protection circuit. While OVP is used to protect the load from a fault in the supply, current limiting is used to protect the supply from a fault (short circuit) in the load. Current-limiting circuitry is included as part of most linear and switching power supplies. Ferros have current limiting as an inherent design characteristic without the additional circuitry required by other regulation techniques.

When considering current-limiting specifications, it is often useful to understand how the limiting is accomplished. Two primary types of current-limiting circuits are commonly found on both linears and switchers, and a third form is unique to switchers. The technique employed can have implications for the system designer.

Constant current limiting performs just what its name implies. When an overcurrent condition occurs (typically 110% to 120% of nominal), the current level is held constant (Figure 7.3). This is usually the simplest and least expensive current-limiting approach. Its disadvantages are that high current levels are present in the system, which can cause potential fire problems if high-current supplies are involved, and that the supply is being continuously operated under 10% to 20% overload conditions, which can place extra stress on all power-handling components. In spite of its relative low cost and simplicity, constant current is not generally found in today's linear or switching power supplies.

Foldback current limiting is the most widely used form of limiting and is found as a standard feature on most linears and switchers. In foldback current limiting, the set point is at 110% to 120% of nominal, as in the case for constant current limiting, but when the foldback circuit is activated, the output current actually decreases (folds back) to some relatively low level (Figure 7.4). In most power supplies, the minimum foldback level is not precisely controlled. A typical specification for foldback current limiting would state, "Short-circuit current

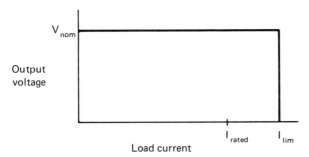

FIGURE 7.3.
Constant-current type of current limiting can place an unacceptable level of stress on power supply components and is not generally used.

shall be limited to a maximum of 40% of full load." This type of current limiting is somewhat more expensive and complex than constant current limiting. These disadvantages are relatively minor when considered in light of the much lower stress levels placed on both the system and power supply components when operating at under 40% of full load current rather than 120%.

Current-limiting circuits are not generally specified in terms of reaction time. But just as in the case of OVP, the type of circuit employed can have a definite impact on the reaction time. The reaction times of constant and foldback current limiting are essentially the same. For a linear, the reaction time of these circuits is more than adequate to protect the supply from just about any fault condition. In the case of a switcher, the same can be said for fault conditions *in the load.* These two current-limiting techniques will not protect a switcher from internal faults such as transformer saturation.

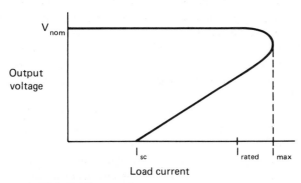

FIGURE 7.4.
Foldback current limiting is the most widely used current limiting technique.

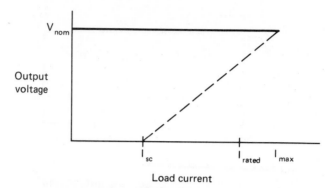

FIGURE 7.5.
Cycle-by-cycle current limiting does not
have a transition area like that of foldback
limiting but goes immediately to a low current level upon activation.

The current-limiting circuits found in most switchers are the foldback
type. They sense an overload condition and then narrow down the pulse width
to fold back the output current level. This circuit must wait for the next half-
cycle of switching frequency to begin narrowing the pulse width. Waiting is fine
if the fault is in the load, but if the fault is a result of transformer core saturation
in the switcher itself, the reaction time of the current-limiting circuit is too long
to prevent the power transistors from blowing.

Cycle-by-cycle, or *instantaneous, current limiting* can protect a switcher from
internal, as well as external, fault conditions. This type of current-limiting
circuit functions not by narrowing the pulse width, but by shutting the switching
transistor(s) off the instant they exceed some maximum level of peak current
(Figure 7.5). Cycle-by-cycle limiting has a faster reaction time than other types
of switcher current limiting since it does not wait for the following half-cycle to
activate. It is activated during the same half-cycle in which the overload occurs.

From the host system's point of view, there is no significant difference
between cycle-by-cycle and foldback current limiting. Both react to a short-
circuit condition by limiting the output current to some small fraction of its full-
load value. Cycle-by-cycle limiting is no more complex or expensive than
foldback limiting, but it can significantly enhance supply reliability by protecting the supply from some internal faults that can happen more quickly than the
activation time needed by foldback limiting.

On multiple-output power supplies, how the current limiting is structured
relative to the various outputs can be just as important as which circuit technique is employed. In most cases, the current-limit circuitry is used to protect
the primary power section and to limit the overall output power of all outputs
combined to some maximum amount. With this approach, it is possible to

lightly load some outputs and greatly overload another to the point of burning it out, without exceeding the overall power rating of the supply. An alternative approach is to have separate current limiting for each output. While the added complexity is not desirable on most multioutput supplies, high-power supplies can benefit from the multiple current-limiting approach since the current levels of the auxiliaries can be large enough to become hazardous under a short-circuit condition.

Power-fail alarm

So far we have discussed protection circuits for the system and power supply hardware. The third commonly used protection circuit was developed to protect the system data. Found in switchers, power-fail alarms are used to warn the host system of impending loss of output voltage. This is a very important protection function of switchers, since there is a much greater likelihood of the ac power line going down than there is of the power supply itself failing.

In many digital processing applications, an orderly system shutdown can only be accomplished if advance warning of loss of power is received. Switchers typically offer holdup times of at least 16 ms. By taking advantage of this holdup capacity and warning the host system sometime prior to the end of the holdup period, it is possible to save valuable data, even in the case of an unplanned loss of input power (Figure 7.6).

Most ac line outages are very short, and the holdup capacity of the switcher can completely mask their occurrence as far as the host system is concerned. During these short interruptions of ac power, there is no need to generate the alarm signal. For that reason, the alarm signal is delayed until some critical time prior to loss of output voltage regulation. The amount of delay is generally variable and is set to give the system the minimum reasonable amount of time required for orderly shutdown. This allows maximum use of the supply's holdup ability while protecting the system from unexpected losses of power.

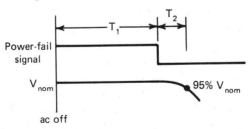

FIGURE 7.6.

The delay between the triggering of the power-fail signal and the loss of output voltage (T_2) can be used to save system software and data.

The circuit that generates the power-fail alarm does not derive its timing by measuring the presence or absence of the ac input voltage; in most instances it senses the voltage present across the energy-storing input capacitors. This allows the circuit to fulfill its mission regardless of whether the ac line goes to zero or just drops below the minimum required operating level (brownout level) for the supply.

The power-fail alarm can also be used at turn on to indicate when the output voltage rises into the regulation band. By measuring the voltage across the input capacitors, the power-fail alarm stays low until the capacitors have been sufficiently charged to support a regulated output voltage. This ability of the power-fail alarm is often used to inhibit system operation at turn on until the output of the supply comes into regulation.

Two areas to watch for when using power-fail alarms are that the signal is not falsely triggered by system load transients, and that the source and sink limits of the alarm circuit are compatible with system requirements. Marginal designs or improperly set alarm thresholds can cause the power-fail alarm to occur as the result of step function increase in load current, which momentarily drains a portion of the charge of the input capacitors. This condition can be corrected by increasing the size of the input capacitors, resetting the threshold trigger point, or reducing the size of the step load changes. A typical power-fail alarm is a TTL-compatible signal that goes to a logic low when the supply's output voltage is about to drop out of regulation due to loss of input power. It is typically capable of sinking a maximum of about 2 mA when in a logic 0 state and of sourcing about 0.4 mA when in a logic 1 state.

DISTRIBUTION

Selection and specification of the correct power supply is only one part of the system designer's problem. A second, and often underappreciated, concern is to effectively get the output of the supply to the exact location of the load.

There are three primary elements of the power distribution problem. The first is to determine where to put the power, that is, in one centralized power supply or smaller distributed ones. Second is to design an adequate grounding and sensing scheme to ensure proper system performance. Third, more and more systems either require more power than is available from a single source or require redundant power supplies, and the question of correctly paralleling power supplies must be addressed.

Power source distribution

At some point in the design of all electronic systems, it becomes necessary to determine the power source requirements and the method of their distribution. Involved in this determination is the weighing of various trade-offs between having a single, centrally located power source and having a number of smaller power sources distributed throughout the system (Table 7.1).

In the case of physically separated, low-power (≤200 W) nodes, it is generally best to design each node with a separate internal power source. When the power needed at a single location is higher, however, say 500 or more watts, the question of power source distribution becomes more complex.

Is a single-kilowatt source better, or should two 500-W units be used? Or four 250-W units? Should the main 5-V power requirement be derived from one source and the auxiliary voltages from another, or should they all be derived from multiple sources? What about on-card dc-to-dc converters?

Six general considerations are involved in the question of power source distribution:

1. Overall packaging and mechanical concept.

2. Required regulation and protective circuitry to individual loads.

3. Reliability factors.

TABLE 7.1
Trade-offs of Power Source Distribution

	Single Power Source + DC to DC	Multiple Power Sources
Packaging and mechanical		
Cooling	Fan	Convection (forced air possible)
System density	High	Medium to low
Power distribution efficiency	Medium (*IR* drop)	High (UL problems possible)
Regulation and protection		
Distribution	Entire system or board level (dc to dc)	Entire system or subsystem (individual supplies)
Reliability		
Power system reliability	High	Medium to low
Sequence control		
Ease of implementation	High	Medium
Noise and stability		
Degree of problems	Medium	Low
Availability		
Number of sources	Medium	High

4. System power-up and power-down sequence control.

5. Noise immunity, power routing, and system ground loops.

6. Market availability of specific wattage ratings.

System power requirements are crucial to packaging concept design. The method of cooling is particularly important. Will the system be cooled by natural convection, forced air, or some form of liquid cooling? Use of natural convection and forced-air cooling will both limit the amount of power that a single source can be expected to provide.

It is unusual to find a convection-cooled power source above 600 W. Forced-air and liquid-cooled units can provide significantly higher power densities, but they may incur reliability penalities. In addition, liquid-cooled power sources are not carried as standard models by power supply manufacturers and would have to be custom designed. How important is high power density to the achievement of the system's design goals?

A second packaging-related consideration has to do with the mechanics of the power routing system. Will a high-voltage, low-current scheme be used to route power to a number of distributed sources, or will low-voltage and high-current from a single source be distributed throughout the system? Use of the former can cause UL safety problems, while the low-voltage, high-current scheme can decrease electrical efficiency due to higher IR drops in the power distribution system.

Regulation and protective circuitry requirements can be important considerations. TTL logic requires 5 V $\pm$ 0.25 V and cannot tolerate a very serious overvoltage condition; therefore, overvoltage protection is essential. But metal oxide semiconductor (MOS) memory requires 12 V $\pm$ 0.6 V and can withstand voltages of 20 V without lasting ill effects.

On-card dc-to-dc converters can be used to provide precise regulation and complete protective functions for critical applications, such as op-amps and D/A or A/D converters on a logic board. An important consideration when using this approach is the need to keep power source duplication to a minimum while still providing adequate protection for the system.

Needless duplication will tend to lower system reliability by increasing the component count. Incorporating on-card dc-to-dc converters means that the loss of a converter can cause the loss of the circuit board. If the design permits individual boards to be easily replaced, system reliability and serviceability could be enhanced. It must be kept in mind, though, that loss of a converter will require the replacement of an entire (increasingly expensive) circuit board.

A second important reliability consideration relates to power technology itself. Linears are rarely used in high-power (over 500-W) applications owing, primarily, to the large bulk and heat penalities associated with them. Switchers, however, represent a relatively new technology.

For example, only a few years ago four switching transistors were needed in 250-W switchers; 1500-W units are now available using four transistors with correspondingly higher reliability. As an increasing number of 1500-W power sources appear on the market, more and more large systems will incorporate on-card dc-to-dc converters supported by a central 5-V source.

System power-up and power-down sequence control can be complicated by using multiple power sources. For example, some IC memory manufacturers require that the −5 V come up first and go down last, relative to the +12 V. When both voltages are derived from a common source, such sequencing is easily accomplished with little or no additional circuitry. With multiple power sources, however, external control circuitry can be required, which will add to the cost and lower the reliability of a system.

Noise immunity, power routing, and the system ground loop are all related factors. Single-point power routing and grounding can significantly reduce or eliminate noise and stability problems; as a result, many designers try to avoid ground loops entirely. When single-point power routing is impractical in light of other system requirements, multiple power sources can help reduce noise and power-routing problems.

Whenever more power is distributed from a single source, whether moving from 60 to 120 W or from 750 to 1500 W, power routing and grounding become important considerations. With increased currents, the IR drop increases and efficiency declines. This lower efficiency can be manifested in poor regulation, oscillating voltages, and generally impaired circuit performance.

When designing a power-routing network, selecting a power supply with the proper current rating often is not enough to ensure proper system performance. One frequently overlooked aspect is system grounding. It is easy to develop a problem in the "ground loop" that occurs when several loads are connected across a single power supply. The result is excessive load interaction.

The simplest power-routing scheme is shown in Figure 7.7. In this scheme, the resistances of the various bus segments (R_b's) convert load currents into voltage drops and cause load interaction. Specifically, these voltage drops

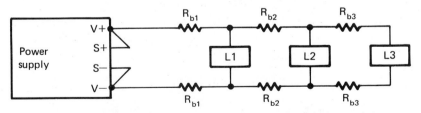

FIGURE 7.7.
This simple power routing scheme can lead to load interaction and less accurate voltage regulation.

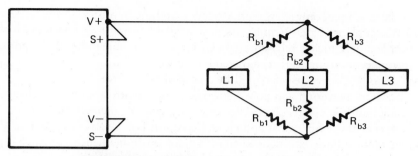

FIGURE 7.8.

In single-point power distribution, load interaction is significantly reduced since each load has a separate bus.

are proportional to the total current drawn by all loads to the right of any given load. This can result in circuit interaction in the form of oscillating voltages or falsely triggered digital circuits.

A better method of routing power to multiple loads is shown in Figure 7.8. The *single-point* distribution technique attempts to isolate each load by running separate power buses to each. This significantly reduces interaction, since now each load forms a complete and independent circuit with the power supply.

Additionally, use of the single-point technique can improve regulation even further for any single critical load. If remote sensing is done across load L1, then a very precisely regulated voltage is seen by that load. Also, if all loads are equal and the R_b's are balanced, all load voltages will be essentially equal to that across load L1.

Unfortunately, single-point power distribution is often impractical due to the additional PC board real estate it requires. Compromises will often be

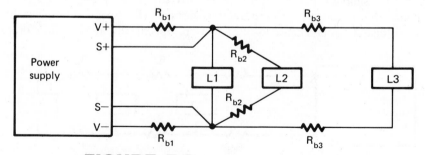

FIGURE 7.9.

Modified single-point power distribution will often provide good results if L1 consumes most of the power and if R_{b2} and R_{b3} are kept small.

successful; for example, many op-amp applications of relatively low frequency and current drain will operate satisfactorily on a distribution network similar to that in Figure 7.9.

In this case, R_{b1} disperses distribution losses to the heavy current load L1, which is remote-sensed. With loads L2 and L3 tied to the same point, R_{b2} and R_{b3} are kept small, and good regulation is obtained across all loads.

Grounding and sensing

Grounding can become even more critical to system performance when multiple voltages are required. In addition to the problems just discussed for each individual output voltage, there is now a chance for interaction between voltages. This is true whether individual power supplies or a single multiple-output supply is used.

The simplest method of connecting three output voltages to three separate loads is shown in Figure 7.10. Although this system of power distribution is easily implemented, it will generally yield poor system performance. The IR drops in the bus network degrade voltage regulation and can also cause interaction between the various output voltages. Use of remote sensing to improve performance can actually cause instability in the regulation owing to the shared distribution resistance (R_b).

A variation of single-point grounding, such as that shown in Figure 7.11, can greatly improve overall power distribution effectiveness. The remote sensing of each output voltage is optional. Its use depends on the current delivered (i.e.,

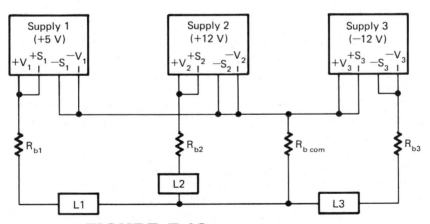

FIGURE 7.10.
Although simple, this multiple voltage power routine scheme can yield poor system performance owing to load interaction and poor regulation.

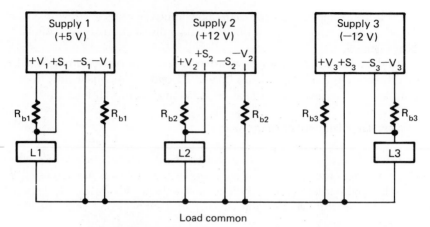

FIGURE 7.11.
As was the case with a single output voltage, multiple voltage systems can realize improved performance when single-point grounding is combined with remote sensing.

the IR drop in the bus) and the level of voltage regulation required by each specific load circuit. In addition, performance can be further stabilized if chassis ground connections are also made at the point and connected, in turn, to the *load common point* with a short conductor.

Remote sensing can be an important aspect of successful power distribution. Simply stated, remote sensing moves the point of regulation from the output terminals of the power supply to the load end of the power distribution bus. It is commonly found on both linears and switchers and is used to improve at-the-load voltage regulation by compensating for the IR drops between the output terminals and the load. The implementation of remote sensing is similar in all power supplies (Figure 7.12).

In both linears and switchers, a voltage-divider network is placed across the output terminals to supply a measurement of the output voltage to an error amplifier, which compares it to a very stable reference voltage and uses their differential to control the regulator. Two things to be aware of when working with remote sense are feedback and derating problems. The error amplifier–control feedback loop has a high gain and is sensitive to the characteristics of the power bus if remote sensing is used. Most power buses are not purely resistive and also have inductive and capacitive characteristics. This can cause improper operation of the feedback mechanism, resulting in either impaired regulation or even voltage oscillation. It is important that these power bus characteristics be recognized and their effect minimized when using remote sensing.

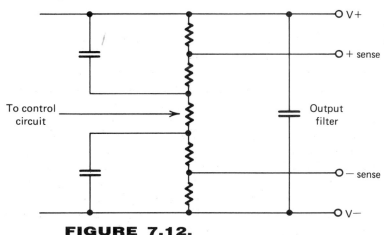

FIGURE 7.12.
In remote sensing, a voltage-divider net-work is used to move the point of regulation from the output terminals to the load.

An often overlooked aspect of remote sensing is the need for power supply derating when remote sensing is employed. Many remote-sense circuits can compensate for up to a 500-mV drop in the power distribution bus. For a 5-V logic supply, operating in remote-sense mode into a bus with 500-mV drop, the load voltage would be 5 V, and the voltage at the supply's output terminals would be 5.5 V, or 10% higher than nominal. The overall output power of the supply has some maximum rated value and is measured at the output terminals, not at the load. If the output voltage is increased by 10% to compensate for drops in the power distribution bus, the maximum load current must be derated by 10% so that the overall power delivered *at the output terminals* does not exceed the supply's ratings.

Paralleling power supplies

Remote sensing, power distribution network design, and internal power supply construction are all important considerations when operating two or more power supplies in parallel. Sooner or later most system designers come across a requirement for more power than can be easily obtained from a single source or for exceptional reliability. In either case, it will be necessary to parallel power supplies to either increase available current or provide redundancy.

There are a number of ways to implement the parallel operation of power supplies, whether linears or switchers. Both internal power supply circuitry and external power busing characteristics affect the success of paralleling. This simplest paralleling scheme is called *direct paralleling* (Figure 7.13). Direct paralleling is used when increased current capability is required. No internal or external

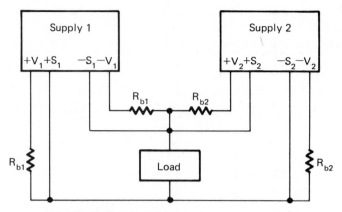

FIGURE 7.13.
Direct paralleling can be implemented with either remote sensing (pictured) or local sensing. Although improving regulation, remote sensing can actually lead to poor load sharing.

load-balancing circuitry is used to implement a direct-paralleling scheme. The cost of this simplicity is loss of regulation and less than perfect load sharing by the supplies. These undesirable effects result from differences in the operating characteristics of "identical" power supplies caused by component and manufacturing tolerance buildups.

Direct paralleling can be implemented using either local or remote sensing. When local sensing is used, the degree of load sharing is primarily dependent on the resistance of the power bus, R_b, between the power supplies and the load. The current delivered by each supply is equal to the voltage drop in the power bus divided by the bus resistance. Therefore, the closer the resistances are, the closer the load sharing. Notice that in this scheme each power supply has its own power distribution bus. If the power supply output terminals are simply jumpered together and then the entire load current carried on a single bus, the load sharing would be significantly impaired. Without the separate bus resistances to provide some balancing, the supply with the higher output would hog the load, delivering the entire load current until it reaches its current-limit point, at which time the second supply would begin delivering the current required in excess of that supplied by the unit in current-limit mode.

A similar result occurs if remote sensing is used with the separate power bus scheme. Remote-sensing directly paralleled power supplies results in poor load sharing. At relatively light (<50% nominal) loads, the power supply with the slightly higher output potential will hog the entire load. As the load increases, that same supply will deliver all the power until it reaches the knee in its current-limit mode. Any increase in the load beyond that level will be delivered by the second supply.

The problem with the remote-sensing form of direct paralleling is that the supply with a slightly higher output voltage can operate at its maximum current level on a continuous basis. If the current-limit set point is left at its usual 110% to 125% of nominal full load, the supply will burn itself out operating continuously in an overload condition. When operating supplies in direct parallel, it is good practice to reset the current-limit maximum level to 90% of nominal to protect it from continuous operation in an overload state and to preserve the reliability of the unit.

There are two ways to get around the choice of good load sharing versus less accurate regulation (local sensing) or poor load sharing versus tight regulation (remote sensing). Direct paralleling can be improved to yield improved load sharing without sacrificing too much regulation by adding some resistance in series with the positive remote-sense lines (Table 7.2). This method of direct paralleling with modified remote sensing will generally yield load sharing roughly the same as the remote-sensing method, while improving regulation by at least a factor of 2 over the local-sensing method. Any power supplies equipped with remote-sensing capability can be employed in the modified remote-sensing technique of direct paralleling.

The second method of improving the operation of power supplies when direct paralleling involves the addition of an internal circuit to each unit and results in excellent load sharing and regulation (Figure 7.14). In this modified direct-parallel scheme, each supply is monitoring its partner's output(s) and adjusting its own accordingly, until a null point is reached in the intersupply

TABLE 7.2
Improving Load Sharing

	+S Resistance, Ω	Typical Load Sharing		Typical Load Regulation, %
		Supply 1, %	Supply 2, %	
Remote sensing[a]	0	5	95	0.20
	2	35	65	0.30
	5	40	60	0.50
	15	45	55	0.75
Local sensing	∞	48	52	2.0

[a]Adding a small resistance to the +S lead of paralleled power supplies will provide load sharing and regulation that is a compromise between the remote and local sensing methods.

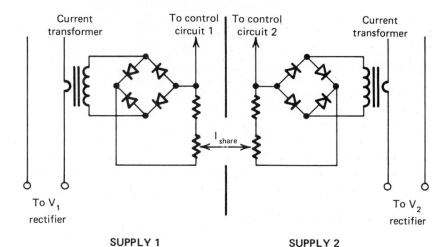

SUPPLY 1   SUPPLY 2

FIGURE 7.14.

Load sharing in directly paralleled supplies can be accomplished by the addition of current-sensing transformers, which are used to monitor the degree of load sharing and provide feedback to the control circuits of each supply until a null point is reached.

monitoring line, at which time the load sharing and regulation will both be very good. Typical performance is load sharing within 10% and load regulation within the specified capability of the supplies. This direct-paralleling scheme is available on some high-current, off-the-shelf power supplies and can be easily incorporated into new designs. Unfortunately, it is not generally easy to add to existing units.

The classical method of achieving good load sharing and regulation is the *master–slave* method. It can be implemented in either of two ways (Figure 7.15). The most generally used method involves external circuitry (Figure 7.15a). In this method, an error amplifier is substituted for the remote-sensing resistor of the slave unit. The error amplifier is connected across the two power supply negative output terminals, and its output is fed into the positive sense input of the slave unit. This method of paralleling uses the resistances R_{b1} and R_{b2} to ensure load sharing just as was the case with the local-sense method of direct paralleling. The regulation, however, is greatly improved since the slave unit tracks the output of the master while the master is remote-sensing and regulating at the load. The alternative approach to master–slave paralleling is to incorporate the required control circuitry into the master unit (Figure 7.15b). For linears, the error amplifier is simply added to the standard circuitry. For switch-

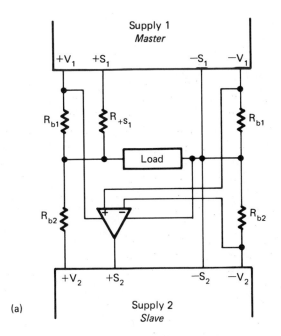

(a)

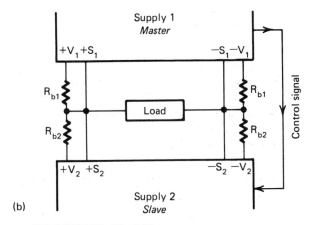

(b)

FIGURE 7.15.

Master–slave paralleling can be achieved using either (a) external or (b) internal control circuitry. In either case, loss of the master means both supplies stop functioning.

ers, two approaches are possible: an error amplifier can be added to the remote-sense circuit as in the case of a linear, or the control circuitry of the master unit can actually be used to provide control for all units.

Whichever master–slave approach is used, the advantages of good load sharing and regulation are not gained without cost. The most obvious cost is the need for additional circuitry. A less obvious disadvantage is a significant reduction in reliability. The additional circuitry has only a minor impact on overall reliability. The major problem arises from the fact that if the master unit fails the entire system is lost. The slave cannot operate without the master; therefore, this approach to paralleling cannot provide redundancy.

For high-reliability situations, such as medical systems or critical communications links, redundancy may be the reason for paralleling units. The paralleling schemes described up to now are designed for systems that require increased power. When redundant power is needed, it is generally necessary to operate the paralleled units in an isolated standby configuration. This can be accomplished by using a modified direct-paralleling scheme that incorporates isolation diodes connected with the outputs of the supplies (Figure 7.16).

Although isolation diodes protect the bus against a low-voltage fault from either power supply, they provide no protection against an overvoltage condition. For the supplies to be truly redundant, some type of OVP must also be provided. The OVP can be either electronic shutdown or crowbar. However, the overvoltage sensing or crowbar must be located on the anode side of the isolation diodes. If it is located on the cathode side, both units will latch off if either OVP fires.

When the diode isolation method of redundant paralleling is used with relatively small loads, there are no real problems. When high currents become involved, however, the picture changes. The *IR* drops across the isolation diodes

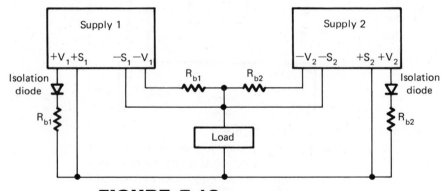

FIGURE 7.16.
The addition of isolation diodes will enable paralled power supplies to operate in an isolated standby parallel configuration.

can be severe at high currents. Even with the relatively low forward drop of Schottky diodes, losses of about 100 W can be experienced with 150-A loads. This level of power dissipation requires additional heat sinking and possibly forced-air cooling, not to mention the expensive high-current diodes and the additional power that the supplies must deliver just to compensate for the diode drops.

An alternative solution, if the application can use a quasi-redundant approach, is to eliminate the isolation diodes and use switchers. Typically, only 20% to 30% of all failures in high-current switchers occur in the output sections. The balance of the failures occur in the primary, or control, sections and result in loss of output. Therefore, it it possible to realize quasi-redundant operation by directly paralleling two switchers, either one of which could handle the entire load.

An important consideration when paralleling to obtain redundancy is the transfer time required for the backup supply to take over the load when the primary unit fails. This aspect of redundant paralleling is related to the degree of load sharing that takes place when both supplies are functioning and the transient response time of the backup unit. Other things being equal, the better the load sharing and the faster the transient response are, the smaller the transfer time.

When redundant supplies are operated with remote sensing without any resistance added to the sensing lines, the regulation will be very good at the load, but load sharing will be almost nonexistent, as discussed under direct-paralleling schemes (Figure 7.13). One unit will hog the load. If the lightly loaded unit fails, the load will see no effect since the other supply was already carrying 90%+ of the load. If the heavily loaded unit fails, however, the result is not as desirable.

Upon failure of the heavily loaded supply, the backup unit will experience an instant step load change of almost no load to full load. The transfer time can range from 50 to 500 ms, which is also the transient recovery time of the supply. When the standby unit sees the initial load demand and attemps to instantly deliver the full-load current, its control circuit will generally go into current-limit mode until the output capacitors have charged. The transient response characteristics of most supplies are designed for fractional (25% to 50%) load changes, not 0% to 100% step function changes.

There are two methods of lowering the transfer time. Both have already been discussed in terms of their effects on regulation and load sharing in direct-paralleling schemes. The first is simply to use local, rather than remote, sensing. The use of local sensing will improve load sharing and reduce the transfer time by reducing the step load change that the backup unit sees when the primary unit fails. The penalty, of course, is less accurate regulation at the load.

The alternative method of reducing the transfer time is to use the *remote sensing through a small resistance* technique of direct paralleling. As seen earlier, this method provides the same approximate load sharing as the local-sensing method, but with much improved load regulation. This technique provides load regulation about midway between the pure local- and remote-sensing methods. Again, the improved load sharing reduces the step load change seen when one unit fails and thereby reduces the transfer time required for the remaining units to pick up the entire load.

Another step that can be taken to reduce the transfer time when a power supply fails is to use three or more units rather than two. If three supplies are paralleled (any one of which can carry the full load), the step load change seen by each of the two surviving units will only be one-half as large as it would be if only two units are paralleled and one fails. This method also had the advantage of running each unit at one-third power, rather than one-half power continuously under normal conditions, and running each at one-half rather than full load upon the failure of one unit. Derating of this magnitude can have a very strong positive impact on power supply reliability. The increased supply reliability, together with the increased redundancy of using three units rather than two, can greatly improve overall system reliability.

SUMMARY

Protection and distribution functions apply to both the power supply and the load. When designing a system power supply and distribution network, much more is involved than merely selecting a power supply large enough to handle the load. The supply can be much more than simply a source of voltage and current. It can provide protection to its own and the system hardware and, in the case of switchers, to the system software and data.

The power distribution network within the system can have a great impact on overall system performance. Remote or local sensing, power distribution, bus resistances, and power supply paralleling techniques all have a major bearing on the quality and reliability of the power delivered to the load. In today's sophisticated electronic systems, it is not enough to have a sophisticated and reliable power supply. The output of the supply, whether protective signals or electric power, must be efficiently and accurately transferred from the supply to the load.

GOVERNMENT AND INDUSTRY STANDARDS

When first confronted with the vast array of national and international safety and electromagnetic interference (EMI) requirements, one can easily find it difficult simply to determine which apply to a particular system. The first step is to break down the regulations into subsets: safety requirements and EMI requirements. Although there is some overlap, these two subsets of national regulations are derived from different international agreements and are enforced differently on a national level.

Before discussing the requirements themselves, it is good to remember what a "standard" is and what it is not. A standard is not a specification. Good specifications are specific and not subject to interpretation. Standards, on the other hand, are guidelines that are interpreted and applied according to local or national customs. The fact that standards and their application can vary with local customs makes the process of identifying the applicable standard for a given system important in avoiding costly overdesign or, equally costly, underdesign and the resulting elimination of the product from the market for failure to meet a particular standard.

INTERNATIONAL
SAFETY AGENCIES

The two most important international safety agencies are the *International Electrotechnical Commission* (IEC) and the *International Commission on Rules for the Approval of Electrical Equipment* (CEE). Their publications take the form of recommendations and are not directly enforceable. Most national standards are based, at least in part, on the recommendations of the IEC and CEE and, in some cases, are in complete harmony with them.

The IEC is headquartered in Geneva, Switzerland, and consists of technical committees made up of representatives from the member countries. Members include most of the industrialized countries of the Western World. The technical committees include experts from manufacturers, users, and national testing laboratories. The recommendations issued by the technical committees are based on a consensus of opinion. In general, the IEC standards meet or exceed the requirements of the member countries; therefore, products designed to meet the IEC standards will often pass the tests adopted by a number of the member countries.

The CEE is a regional, European agency. It is one of the world's most powerful standards-making organizations since most European countries have adopted large portions of the CEE standards as national requirements. Membership in the CEE is strictly European; the United States participates only as an observer. The United States is an active member of the IEC, however, and many of the CEE standards have been totally adopted by the IEC.

Neither the IEC or CEE have specific standards that apply solely to power supplies. Rather, the power supply must meet the standard that applies to the end product or system in which it is installed. This same situation exists when European national-level standards are considered. Only the UL and CSA have specific standards for power supplies. But even in the case of UL and CSA, the power supply is expected to meet further requirements as defined for the end product or system. As a result, few power supply manufacturers list compliance with UL 1012 on their specifications. Instead they list UL 478. The power supply standard is UL 1012, but UL 478 is the specific standard for the electronic data-processing systems in which many power supplies are typically used.

Power supplies hold an important spot in any electronic system. They interface between the high-voltage (high-hazard) ac input line and the typically low-voltage (low-hazard) working circuits of the system. As a result, many system-level minimum safety standards focus on the power supply within the system to ensure safe operation. Another result is that while the system requirements themselves might vary quite a bit, from office machines to medical equipment to data-processing systems, the power supply requirements vary to a much lesser extent. There is more variation in power supply requirements between agencies than there is between system requirements within a given agency.

On a national level there are many testing agencies to consider. A partial listing includes Underwriters Laboratories (UL) in the United States, Canadian Standards Association (CSA) in Canada, Verband Deutscher Electrotechniker (VDE) in West Germany, Svenska Elektriska Materielkontrollanstalten (SEMKO) in Sweden, Schweizerischer Elektrotechniker Verein (SEV) in Switzerland, and Standards Association of Australia (SAA) in Australia. Many other countries have safety standards but will accept either UL or VDE approval. In general, a system that meets UL, VDE, and IEC standards will also meet most other national safety requirements. As a result, a good knowledge of the requirements of UL, VDE, and IEC will satisfy most practical situations.

MAJOR SAFETY STANDARDS

The following key safety standards are important to understand when designing products for distribution in the major international markets. Problems of compliance with all these standards at once arise as a result of the leakage current, creepage distance, and dielectric withstand (isolation) test requirements of the standards (Table 8.1).

TABLE 8.1
International Safety Requirements

Safety Standards	Leakage Current, Line to Ground, mA	Creepage Distance, Live Parts to Dead Metal, in. (mm)	Dielectric Withstand Voltage (rms)	
			Inputs to Ground, Vac	Inputs to Output(s), Vac
UL 114	0.25	0.094 (2.4)	1000	1000
UL 478	5.00	0.094 (2.4)	1000	1000
CSA 22.2 No. 143	5.00	0.094 (2.4)	1000	1000
CSA 22.2 No. 154	5.00	0.094 (2.4)	1000	1000
IEC 380 Class I	3.50	0.120 (3.0)	1250	3750
IEC 380 Class II	0.25	0.120 (3.0)	3750	3750
IEC 435 Class I	3.50	0.120 (3.0)	1250	3750
IEC 435 Class II	0.50	0.120 (3.0)	3750	3750
VDE 0804 Class I	3.50	0.120 (3.0)	1500	2500
VDE 0804 Class II	0.05	0.120 (3.0)	2500	2500
VDE 0806 Class I	3.50	0.120 (3.0)	1250	3750
VDE 0806 Class II	0.25	0.120 (3.0)	3750	3750

International

European Computer Manufacturers Association (EMCA)

EMCA57: *Safety Requirements for Data Processing Equipment.* Applies to data processing and data communications equipment used in an office or industrial environment. This strictly European standard is essentially the same as IEC 435.

International Commission on Rules for the Approval of Electrical Equipment (CEE)

CEE10 Part 1: *Specifications for Electric Motor Operated Appliances for Domestic and Similar Purposes.* Applies to domestic electric appliances such as clothes washing machines, air conditioners, and the like. Like IEC 335, it is not applicable to electronic equipment used in business or industrial environments.

CEE10 Part 2P: *Particular Specifications for Business Machines.* Applies to business machines such as typewriters and copiers used in an office environment. Similar to IEC 380.

International Electrotechnical Commission (IEC)

IEC 335: *Safety of Household and Similar Electric Appliances.* Like CEE10 Part 1, it does not apply to electronic equipment operated in business or industrial environments.

IEC 380: *Electrical Safety of Office Machines.* Applies to business machines operated in an office environment. Similar to CEE10 Part 2P.

IEC 435: *Safety of Data Processing Equipment.* Applies to data-processing and data-communications equipment. Similar to ECMA57.

National

Canada: Canadian Standards Association (CSA)

CSA 2.2 No. 143: *Office Machines.* Applies to business machines operated in an office environment. Similar to UL 114.

CSA 22.2 No. 154: *Data Processing Equipment.* Applies to equipment designed to prepare, store, process, display, or transmit data. Similar to UL 478.

Germany: Verband Deutscher Electrotechniker (VDE)

VDE 0730 Part 1: *Specifications for Electric Motor Operated Appliances for Domestic and Similar Purposes.* Does not apply to electronic equipment operated in business or industrial environments. Similar to CEE10 Part 1 and IEC 335.

VDE 0730 Part 2P: *Particular Regulations for Office Machines.* Applies to business machines operated in an office environment. Similar to CEE10 Part 2P and IEC 380. To be replaced by 0806.

VDE 0804: *Telecommunications and Data Processing Equipment.* Applies to all EDP and telecommunications equipment. This specification was the original basis for IEC 435. To be replaced by 0805.

VDE 0805: Updated and clarified version of 0804; no major changes from 0804.

VDE 0806: Updated and clarified version of 0730; no major changes from 0730.

Japan: Electrical Appliance and Material Control Law (EAMCL)

EAMCL: This standard covers all electric and electronic systems and machines marketed in Japan. It is similar to the combined requirements of IEC 380 and IEC 435.

United Kingdom: British Standards (BS)

BS 3861: *Electrical Safety of Office Machines.* Applies to business machines operated in an office environment. Similar to CEE10 Part 2P and IEC 380.

BS 4644: *Recommendations for Safety of Office Machines and Data Processing Equipment.* Similar to IEC 435.

United States of America: Underwriters Laboratories (UL)

UL 94: *Plastic Materials for Parts in Devices and Appliances.* Covers the flammability of plastic materials used in domestic, office, and industrial environments. There is no IEC or European equivalent.

UL 114: *Standard for Office Appliances and Business Equipment.* Covers all business machines operated in an office environment. Similar to CSA 22.2 No. 143.

UL 478: *Standard for Electronic Data Processing Units and Systems.* Applies to all data-processing equipment used in office or industrial environments. Similar to CSA 22.2 No. 154.

UL 1012: *Power Supplies.* Covers all commercial and industrial power supplies. There is no IEC or European equivalent.

IEC, UL, AND VDE SAFETY REQUIREMENTS

The enforcement powers of these agencies vary widely. The IEC has no enforcement powers; it only publishes recommended standards. However, since most of its recommendations are ultimately adopted by the national agencies that do have enforcement abilities, the IEC recommendations must be taken very seriously. Underwriters Laboratories is not a government body; it is a private, nonprofit corporation that works on a voluntary basis with the U.S. electronics industry to develop safety standards for commercial and consumer products. Although without its own enforcement powers, many state and local governments have adopted UL standards as legal requirements. The VDE has the most enforcement power among these three agencies.

The West German safety laws have made the VDE standards an internationally recognized electrical safety specification. The legal requirement is that all electronic systems must be designed and manufactured to conform with the VDE standards. Formal testing is not a legal requirement. However, any product or system found not to comply can be banned from distribution. Certification by VDE is sought for almost all consumer goods and appliances and for commercial electronics built by importers. Certification is not usually obtained by German manufacturers of commercial and industrial electronics but is a strong marketing tool for importers.

The categories of standards of most interest are Data Processing Equipment, UL 478, CSA 22.2 #154, VDE 0804, and IEC 435; and Business Machines, UL 114, CSA 22.2 #143, VDE 0806, and IEC 380. In the case of power supplies, the requirements are almost identical (i.e., UL 478 treats power supplies basically the same as UL 114, VED 0804 is similar to VDE 0806, etc.). Other categories of standards, such as medical or communications, vary only slightly with regards to specific application concerns such as leakage current in medical equipment or shielding in communications equipment.

Both the IEC and VDE recognize four categories of insulation: basic or functional insulation, which provides minimum protection from shock; supplementary insulation, which provides backup protection if the basic insulation fails; double insulation, a combination of basic and supplementary insulation; and reinforced insulation, which is improved basic insulation that provides the same protection against shock as double insulation. These insulations are not required to be a solid material but may consist, in part or whole, of creepage and clearance spacings. The IEC and VDE use the term creepage to define what UL terms "over-surface spacing"; clearance corresponds to what UL terms "through-air spacing" (Figure 8.1).

The IEC classifies the voltages present in electronic equipment as hazardous, extra low, and safety extra low when specifying the insulation requirements. A hazardous voltage is any voltage above 42.4 V peak. An extra low voltage (ELV) is considered present when the nominal voltage is below 42.4 V peak but is not adequately isolated from a hazardous voltage and, as a result, must be treated as a hazardous voltage. A safety extra low voltage (SELV) is any voltage below 42.4 V peak that is either obtained from an energy-limited source (200 VA) or is protected from any conceivable reach-through of a hazardous voltage.

The VDE requirements are identical to the IEC except in the area of SELV transformers, which are required to have more insulation by the VDE. The UL requirements are much less strict. For example, the minimum spacing requirement for UL is 0.1 in. (2.4 mm), whereas VDE and IEC state a 3-mm minimum. A result of these variations in design criteria can be seen in the production testing criteria of these agencies. The IEC and VDE require 100% testing of the isolation between primary and ground (1500 rms for 6 s), but not between primary and secondary due to the high isolation inherent in the construction requirements (Figure 8.2). The UL, on the other hand, requires both primary to ground (1000 V rms for 5 s) and primary to secondary (same test) because of the less rigorous construction and design requirements.

For power supplies, especially switchers, the IEC and VDE requirements can be tough to meet. In a linear, it is often enough to replace the UL-type

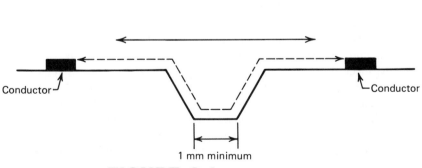

FIGURE 8.1.
Creepage and clearance distances.

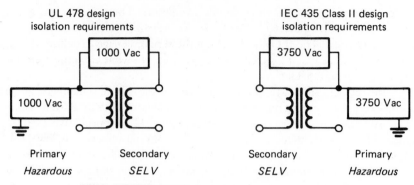

FIGURE 8.2.
IEC and VDE production testing requirements are less rigid than UL and CSA owing
to their stricter design requirements.

power transformer with an IEC/VDE transformer and make some relatively
minor component or construction changes to transform a UL-only model into
an interational model. Switchers, with their much larger primary circuit and
feedback from secondary into the primary, pose a much more complex problem.
It is quite often true that a switcher designed solely for UL cannot hope to meet
IEC/VDE without a major redesign.

The IEC/VDE isolation requirements can be especially difficult to meet
when it comes to switcher magnetics, particularly the power transformer that
has a high ratio of primary to secondary turns. Because of the relatively small
number of turns on the secondary winding, close proximity is essential to ensure
good coupling and efficiency. This is contrary to the IEC/VDE desire for high
secondary to primary isolation. If attacked as an afterthought, this dilemma can
lead to large and expensive transformers. However, if considered at an early
design stage, an IEC/VDE switcher need not cost any more than a switcher that
does not meet the European standards.

Another area in which there can be problems in redesigning a UL system
into a UL, IEC, VDE system is leakage current. If designed around UL 478 and
its 5-mA limit on leakage current from line to ground, it can be difficult to
design to meet the IEC/VDE 3.5-mA limit. However, there is hope if the
equipment was designed for UL 114 (business machines), since its 0.25-mA
leakage requirement is the same as IEC 380 and VDE 0806. It should be noted
that the IEC/VDE leakage requirement depends on the application of the equipment. Portable units (units small enough to be carried) must meet the 0.25-mA
requirement, while stationary equipment must meet 3.5 mA. UL does not make
this distinction.

So far, UL sounds easier to meet than either IEC or VDE. Is it true that
U.S. manufacturers will have a difficult time making equipment "safe" enough

for the European market, while the European manufacturers can simply send their standard products to attack the U.S. market? Not exactly.

One important area in which UL is definitely ahead of both the IEC and VDE is flammability standards. Neither the IEC nor the VDE, at this time, has such standards. The United States has been attempting to have the IEC adopt the UL flammability standards but has had no success up to now. The flammability standards of UL 94, for instance, are a totally unfamiliar area to European manufacturers attempting to enter the U.S. market for the first time. The result is that, until the standards are made compatible on an international basis, probably through the IEC, both U.S. and European electronics manufacturers can have significant problems when attempting to sell their product outside their respective home markets.

The IEC/VDE temperature and abnormal testing differs from UL. All require a 65°C temperature rise, but the IEC/VDE tests both primary and secondary, whereas the UL considers only the primary side. Abnormal testing is treated in a similar manner by both agencies. In abnormal testing, all components *will be* opened or shorted and the unit checked for unsafe temperature or electrical isolation conditions.

Labeling requirements are an area in which almost no agreement exists. Most countries, quite understandably, want the labeling in their national language. Some, such as Canada, require bilingual labels. IEC/VDE require that power supplies be labeled for output wattage (not a VA rating), whereas UL requires the label state the input power in VA (not watts). In addition, IEC/VDE require that input voltage and grounding connectors be labeled with particular "international" signs, which the UL does not employ.

Enforcement powers

The safety standards of VDE are highly regarded and have the force of law in West Germany. In fact, the VDE safety standards were used as the basis for the IEC safety standards. The VDE itself is the National Association of German Electrical Engineers. The association develops the standards, which then become part of the West German safety law.

There are four types of safety approvals possible under West German law:

1. CB *certificate:* For consumer and commercial products that do not contain electronic components.

2. *Certificate of conformity:* Applies only to the specific unit submitted for testing. This procedure tells the applicant whether the product will conform to the required safety standards. A certificate of conformity cannot be advertised or published in product literature.

3. *Certificate of conformity in conjunction with ongoing factory surveillance:* For components or subsystems (e.g., power supplies) that fully comply

with required safety standards when installed in the end system or machine.

4. *VDE-tested safety (GS) mark for end-user equipment:* The VDE-GS mark is Europe's most widely recognized safety approval for consumer appliances. It is not used for components or subsystems such as power supplies.

The VDE-GS mark is not a mandatory requirement for systems imported into West Germany. It is only necessary that products be designed and manufactured to comply with the VDE standards. Noncompliance, however, can lead to being banned from distribution, so many manufacturers have their products tested to avoid any possible misunderstandings or problems once they begin marketing their products. The VDE-GS mark itself is not always sought by manufacturers of commercial or industrial equipment since it adds little to the marketing effort, but, typically, costs $10,000 to $25,000 and can take 6 to 18 months to obtain. It is a rare consumer product, however, that is marketed without the VDE-GS mark of approval.

The enforcement powers of UL and CSA differ substantially from those of VDE. Where the VDE has the force of law and products can be banned from distribution for noncompliance, the safety standards of UL and CSA are primarily voluntary standards. Some local governments (e.g., states or cities) have adopted UL or CSA as legal requirements. The enforcement differences between VDE, UL, and CSA can be seen in the implementation of the standards:

1. VDE *approves* systems or machines, but not components or subsystems such as power supplies.

2. UL *recognizes* components and subsystems such as power supplies.

3. UL *lists* systems or machines that are designed for end-use applications.

4. UL *classifies* materials such as plastics in terms of flammability and other properties.

5. CSA *certifies* systems and components, including power supplies.

Although without the legal force of VDE, approval, recognition, listing, and classification by UL and certification by CSA have been adopted by industry so completely that the market for noncompliant products is severely limited.

ELECTROMAGNETIC INTERFERENCE

Prior to the widespread use of switchers, electromagnetic interference (EMI) was almost a nonentity as far as power supplies were concerned. Electromagnetic interference consists of the unintentional generation of either radiated or conducted energy that interferes with the normal operation of other electronic systems, most typically radio communications equipment. Linear or ferroresonant power supplies generally operate at 60 Hz and produce little or no EMI. It is possible for the power transformer in a linear or ferro supply to interfere with the proper functioning of a cathode-ray tube display, but this is more of an internal system problem than an EMI problem.

Switchers, on the other hand, and the digital systems that they power, are capable of producing significant levels of EMI. The concern is not that the EMI levels of a properly designed switcher interfere with the functioning of the host system; the concern with EMI arises from the fact the system as a whole, including the switcher and digital circuitry, may interfere with the functioning of *other* systems. It is the desire to reduce the potential for intersystem EMI problems that is the basis for all major EMI regulations.

The technical foundations for all EMI limiting regulations are strikingly similar because they are all based on the natural electromagnetic spectrum. As a result, the various national EMI regulations vary more with regard to the specific classes of equipment regulated than in terms of the EMI limits themselves.

Limits for incidental interference have existed for quite some time on an international level and are based on treaties intended to ensure that adequate radio communications reception is possible at a given distance from any specific class of EMI sources. This ensures that products manufactured in one country can be readily used in another country without causing EMI problems.

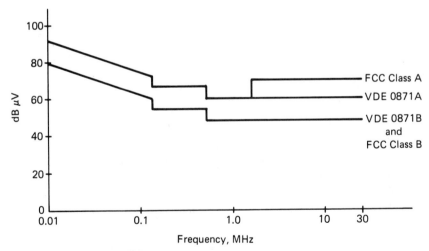

FIGURE 8.3.
The U.S. and West German EMI limits are similar; although the U.S. (FCC) limits are currently less strict, it is expected that in the future they will become identical with the stricter West German EMI limits.

In the United States, however, until recently only equipment for export to Europe or for sensitive military applications had to meet EMI limitations, typically MIL-STD-461 or VDE 0871. As a result of the growing number of complaints that the FCC has received about EMI generated by digital systems, a new EMI specification has been imposed on U.S. manufacturers (Figure 8.3).

EMI limits

The FCC regulations cover conducted and radiated EMI from electronic computing devices. A computing device is defined as any electronic device or system that intentionally generates and uses radio-frequency energy in excess of 10 kHz. The reasons cited by the FCC for the increase in EMI problems include the following:

(1) Digital equipment have become prolific throughout our society and are now being sold for use in the home; (2) digital technology has increased the speed of computers to the point where the computer designer is now working with radio frequency and electromagnetic interference (emi) problems—something he didn't have to contend with 15 years ago; (3) modern production economics have replaced the steel cabinets which shield or reduce radiated emanation with plastic cabinets which provide little or no shielding.

These new regulations apply to all digital systems intended for use in the United States. This affects virtually all computers and their peripherals, process controllers, medical equipment, personal computers, video games, and so on. The FCC has actually established two different EMI limits depending on the type of equipment involved. Devices intended for use in a home or residential environment must meet the relatively strict class B requirements. Commercial, industrial, and business systems must meet the less difficult class A limitations.

It is as important that a system designer understand what the FCC and VDE do not require as it is to understand what is required. Both agencies have *system*-level requirements, not subsystem or component requirements. Technically, power supplies do not have to comply with the EMI limits to be marketed. As a practical matter, however, any noise that the power supply introduces into the system must be cleaned up as part of the overall system's EMI housekeeping.

While the FCC and VDE specifications are similar in that they both apply only to complete systems, they are dissimilar when it comes to determining which system must meet which requirement. The FCC differentiates systems based on their operating environment, whereas the VDE uses operating frequency as its decision parameter. In the United States it was possible for the FCC to segment digital systems into those that operate in "friendly," EMI-tolerant, commercial or industrial environments (class A) and those that operate in "unfriendly," EMI-sensitive, residential environments (class B).

In Europe, however, there is less distinction between industrial and residential areas. The relatively homogeneous nature of the European infrastructure mandated that the VDE create two equipment classes based on operating frequency. Systems operating above 10 kHz are presumed to have a relatively high potential for producing EMI and must meet the stricter regulation, VDE 0871. Those systems operating at or below 10 kHz are required to comply with the less restrictive VDE 0875 limits.

The VDE has gone one step beyond the FCC, further dividing each specification into multiple categories: 0871A, B, and C; and 0875N and N-12. Compliance with the lower limit (0871B or 0875N-12) confers "general approval." The advantage is that individual production units do not need to be tested for EMI. Systems that meet the less strict limits receive only "single approval" and must be tested on a unit by unit basis as they are produced.

In addition to individual production testing, systems that meet VDE 0871A or C cannot be operated without an *individual permit* for each installation site. Equipment that meets only VDE 0871C must also be tested at the installation site, not on the production line. All in all, for digital systems intended for sale in the German or European markets, it makes a great deal of sense to design for VDE 0871B.

The testing and verification requirements of the FCC are less strict. Each manufacturer is expected to take the necessary design and production steps to ensure that his products comply with the appropriate EMI limits. Submittal of a typical unit or copies of the test data are not generally required. The FCC reserves the right to request such proof at any time if it receives complaints and suspects that a particular piece of equipment is a potential source of interference.

Additional requirements of the FCC are the label and operation manual requirements. The label on a device or system must be placed in a conspicuous location with specific warning about the interference potential and the operator's responsibility when operating the equipment in a residential area. The operation manual must provide a similar warning and list some simple measures that the user can take to eliminate any interference problems that might arise.

Prior to the current interest in the FCC and VDE EMI limits, MIL-STD-461 was often used as the bench mark when specifying EMI requirements. It is still one of the most commonly cited military EMI specifications. As indicated by its full title, *Electromagnetic Emission and Susceptibility Requirements for the Control of Electromagnetic Interference*, MIL-STD-461 takes the problem of EMI one step farther than either the FCC or VDE and concerns itself with susceptibility as well as emissions.

The results are five primary categories, each subdivided into numerous subsections. The five primary categories are conducted emissions, conducted susceptibility, radiated emissions, radiated susceptibility, and unique requirements for general-purpose equipment. The specific requirements for a particular piece of equipment are typically only a subset of the overall standard. Most

commercial equipment is considered to fall in the category of "unique requirements for general-purpose equipment." This section of the standard is generally
interpreted as less stringent than either the FCC or VDE limits.

Keeping the electromagnetic environment clean

Conducted EMI is the cause of most EMI problems, including radiated
EMI. The system designer's primary concern used to be that extraneous RF
energy conducted into a digital system could interfere with the proper functioning of the system. While that problem still exists, the emphasis has shifted to the
need to meet the FCC and VDE EMI requirements. The reason for the shift is
that the conducted RF energy generated by the majority of commercial switching power supplies currently on the market is filtered enough to prevent interference with all but the most sensitive digital circuits, but it still exceeds the
FCC and VDE limits (Figure 8.4).

The two forms in which conducted EMI is manifested are differential
mode and common mode. *Differential-mode* EMI occurs as an electrical potential
across the ac lines; hence it is sometimes called *line-to-line* interference. The
differential mode currents in the ac lines are 180° out of phase, causing the
corresponding electromagnetic fields to cancel. The result is that differential-
mode EMI is attenuated very quickly and is not generally a significant problem.

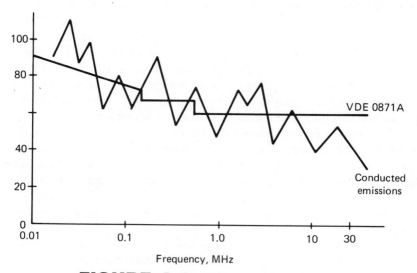

FIGURE 8.4.
Typical conducted emissions for switching
power supply with a "standard" EMI filter.

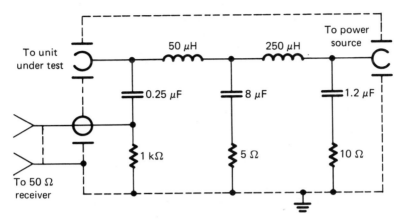

FIGURE 8.5.

Recommended line impedance stabilization network (LISN) as defined by the FCC and VDE.

Most conducted EMI is of the *common-mode* type and appears on all ac lines with respect to ground. As a result, it is readily distributed through the ac line and can cause a substantial amount of radiated EMI. Meeting the EMI requirements of the FCC or VDE requires careful shielding for reduction of radiated EMI and the use of ac line filters for reduction of conducted EMI.

Conducted EMI, both common and differential modes, is generally controlled by the use of ac line filters. The attenuation of such filters depends on a number of factors, including ac line and equipment impedances and the number of filter sections. Because of the difficulties in predicting ac line and equipment impedances, no direct, quantitative method is available when designing effective EMI filters.

A 50-Ω impedance is generally used as an approximation of ac line impedances throughout the world. In fact, the FCC and VDE both define the line impedance stabilization network (LISN) similarly, using a 50-Ω impedance (Figure 8.5). Both groups specify a quasi-peak detector or correlated equivalent for measurement. A significant difference, however, is the specified receiver bandwidth. The FCC's 1-kHz bandwidth in the lower frequencies is actually a tougher test than the VDE specification of a 0.2-kHz bandwidth between 10 and 150 kHz.

A more easily controlled parameter that affects filter performance is the number of filter sections (Figure 8.6). The higher the number of sections is, the greater the filtering action. The simplest type, single-section filters generally yield no more than 55-dB attenuation above 3 MHz. At around 150 kHz, attenuation drops to the 15- to 22-dB range. Single-section filters are not usually

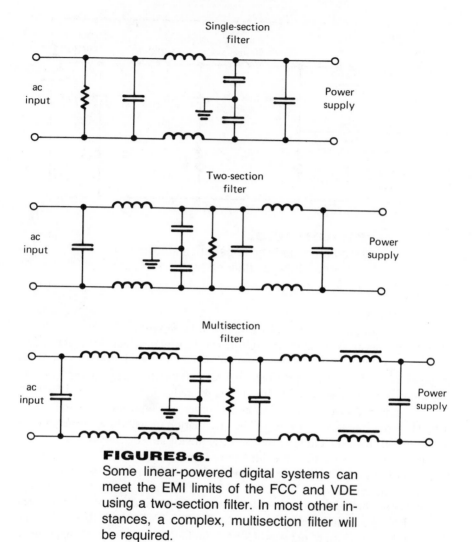

FIGURE8.6.
Some linear-powered digital systems can meet the EMI limits of the FCC and VDE using a two-section filter. In most other instances, a complex, multisection filter will be required.

successful in meeting even the most lenient EMI (FCC class A) requirements, even when the system is powered by a linear.

Typical digital systems powered by linear power supplies require at least a two-section filter. This filter configuration can provide 70 dB of attenuation of frequencies as low as 500 kHz, a definite improvement over the single-section configuration. Some systems, even when powered by linears, are so noisy that two-section filters are not adequate. In these cases, complex multisection filters must be employed.

Systems powered by switchers are obvious candidates for complex filters. Adding a poorly filtered switcher to a system that already demands a multisection filter can cause serious EMI problems. Power supply manufacturers are beginning to recognize the seriousness of the new FCC regulations and are now producing some units that meet the FCC and VDE EMI limits. While not specifically required to meet the limits, since it is classified as a component, a switcher that satisfies the international EMI regulations can certainly make the system engineer's job much easier.

As can be seen, the problem of conducted EMI is not simple. Even when a linear supply is used, complex filtering is often necessary. When using a switcher that itself meets the international EMI regulations, the problem is of a similar magnitude as when employing a linear. In the final analysis, the designer of a digital system is faced with the empirical task of developing an EMI filter network with small losses at 60 Hz, very high losses in the 10-kHz to 30-MHz range, and impedances that closely match the ac line and the system.

The problem of radiated EMI can be split into two subsets. Radiated EMI coming from an enclosed switcher is not a serious problem. Enclosed switchers are, typically, medium- to high-power units and are used in relatively large systems. The aluminum–steel case of the switcher along with the (typically) steel enclosure of the system provides more than adequate shielding. The only remaining concern is the control of conducted EMI.

Open-frame, low- to medium-power switchers can present the system designer with more serious radiated EMI problems. Obviously, there is no enclosure to contain the power supply's EMI, and to add to the problem, more and more of the systems employing open-frame switchers are themselves housed in plastic enclosures. The potential exists of having a 400- 500-W open-frame switcher operating in an almost completely unshielded environment.

In the case of low- to medium-power open-frame switchers, it is generally necessary to provide some type of EMI shielding as part of the system enclosure. Typically, either conductive composites or coated nonconductive plastics will be used for the enclosure. Conductive composites are a relatively new technology. Problems can arise in the mechanical strength of the enclosure, the distribution of filler particles over complex enclosure geometries, and the environmental tolerance of the conductive composites. When employed under the correct circumstances, they can provide a cost-effective approach to EMI shielding.

When coated nonconductive plastics are employed, all ribs and bosses should have an adequate radius on all edges, deep holes should be avoided, and internal walls should not meet at sharp right angles. If not properly designed, the coatings can build up in corners and become thin along edges. Cracks in the coating along an edge can, in fact, act as an antenna and increase instead of decrease radiated EMI problems.

In either case, the time to reduce radiated EMI is when the system is at the circuit design and PCB layout stage. Add grounded metallic covers over emitting devices, keep electrical paths as short as possible, ground all shielded cables, use a single-point ground scheme, and do not use shields as return lines. In short, be conscious of potential EMI problems and follow good design practices.

SUMMARY

When determining which safety and EMI standards will be specified for a particular application, it is well to remember that all such standards apply to machines, in the case of safety, and to systems, in the case of EMI, not specifically to power supplies. In fact, some compromises can often be necessary. The safety requirements of high isolation and low leakage can be somewhat at odds with the need for filtering and grounding to control EMI.

In addition to the problem of integrating safety and EMI standards with each other, there can be a problem integrating them into the power supply specification itself. The safety requirements of IEC/VDE can lead to larger package sizes, lower power densities, and less efficient magnetics, especially in the case of switchers. Again, for switchers, the filtering required to meet the EMI requirements of the FCC and VDE can actually slow the transient response time of the unit.

Finally, it must be pointed out that this chapter has dealt only with data processing equipment and business machines. The standards for medical or dental equipment are somewhat different and can vary widely from country to country. Even the international standards multiply. In most countries, the same agency that oversees safety standards also oversees EMI limits. The United States is the only major exception, with UL responsible for safety standards and the FCC responsible for EMI limits.

As international trade continues to increase in importance, compliance with the safety and EMI regulations will continue to be a growing concern in the design and production of electronic systems. It is important to have a good basic understanding of the differences and parallels between the various national standards in order to intelligently specify the most realistic parameters that will ensure compliance at a reasonable cost, both in monetary terms and in terms of the technical trade-offs involved.

MAKE OR BUY
DETERMINATIONS

An electronics company that continually un-
dertakes a program of uncontrolled vertical integration can be likened to a juggler who keeps
more and more balls in the air until he loses
control of all of them. Successful companies
place a great deal of importance on make-or-buy determinations. In most major
corporations, the make-or-buy decision for subassemblies, such as power supplies, is made at two different levels of management.

Initially, upper management generates a statement of policy outlining the
corporation's general philosophy in the area of make or buy. Specific make-or-
buy decisions are then analyzed at the middle-management level, and the specific decisions are monitored to assure conformance to that policy.

STATEMENT OF POLICY

A typical make-or-buy policy statement might be as follows:

> The make-or-buy program analysis should be confined to important items which,
> because of their complexity, quantity, cost, or overall system performance impact,
> require management review

As a general guideline, the make-or-buy analysis should not include items or work effort costing less than 1% of the total system cost, or $500,000, whichever is less, and should not include raw materials or off-the-shelf items.

Other (often unstated) policy factors can strongly influence the attitude of middle management. The make-or-buy analysis can be biased toward the make decision if upper management is interested in developing a diversified production capability, if they feel the design flexibility can be increased by bringing as much in-house as possible, or if they feel the return on investment looks (or can be made to look) attractive to stockholders.

Diversified production

Diversified production capability sounds like a worthwhile goal, since a broad-based company would seem likely to have increased stability. This is probably true if the products of the company are sold in the competitive market. However, if they are used strictly or primarily for internal consumption, inflexibility in the use of materials, parts, or supplies often results.

A purchasing firm can buy from any source that offers the best combination of price, quality, and service. It is free to substitute items, shift from one source to another, or split orders among competing sources as terms and conditions warrant. This freedom can be greatly restricted for a firm whose management is committed to procurement by manufacture.

Companies that produce their own material as a sideline can rarely afford sufficient research and development investments to keep abreast, much less ahead of, firms for which such products are a primary business. It is likely to be only a matter of time until the company producing such a needed article as a sideline falls behind the major producer of the article in quality, performance, and cost improvements.

If the quality of the sideline component is a factor in the performance of the primary system the company is producing, the results of making, instead of buying, can well be disastrous.

Design flexibility

Increased design flexibility seems to be a realistic result of increasing the proportion of items manufactured in-house. However, overall system design can actually be less flexible owing to a dilution of effort.

Manufacturing an item as a sideline often involves new equipment, new skills, new technology, and new personnel. Moreover, every time another somewhat unrelated production unit is added to the organization, there is some loss of cohesion and unity in management, which is certain to produce a new set of technical and administrative problems.

Administrative and technical expertise is a limited resource which can be easily overloaded as additional items or processes are integrated into a company's activities.

Return on investment

Attractive return on investment is a common goal of all businesses. The judgment of what constitutes a proper return on invested capital is difficult to make. Make-or-buy decisions are often made without studying alternative uses of capital. In general, investments should be made based upon the minimum acceptable return a company expects to make in its field of expertise.

An investment that yields 25% per year sounds pretty good. However, if a company has been doubling its invested capital every two years, it sounds horrible!

A complete cost analysis in a make-or-buy situation is very complex. It includes factors such as the following:

1. Normal direct manufacturing costs.

2. Normal overhead expenses (which vary from industry to industry).

3. Individual procurement of components versus procurement of the whole article.

4. Absorption of a temporary reduction in work load to retain specialized skills and organizational capacities.

5. Utilization of existing skills and facilities.

6. Placement of work at sources capable of producing follow-on requirements.

A more detailed discussion of costs will follow later, but it is obvious from this listing that there are many less than obvious costs that can turn an in-house production effort into a financial disaster.

The impact of a biased policy can be seen in the following study of the experience of a medium-sized manufacturer of minicomputers. Upper management felt that the in-house production of power supplies gave them increased flexibility and an attractive return on investment. As long as linear-type power supplies were designed into their systems, no serious problems were experienced. However, their first venture into switchers led to some unexpected difficulties (Figure 9.1).

As had previously been done with linears, the design of the switching power supply was assigned to a junior member of the engineering team. Serious problems in this design effort soon became evident, so the switcher project was moved up the ladder to a more senior engineer. After being moved up the ladder

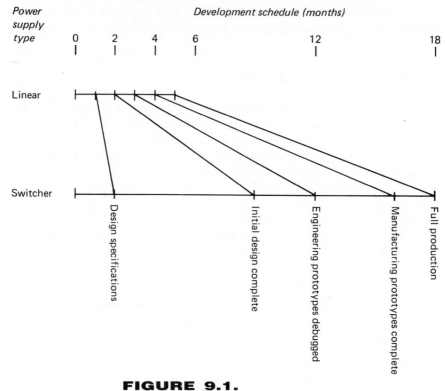

FIGURE 9.1.
Typical power supply development schedules.

a few more times, the project finally ended up on the desk of the vice-president òf engineering!

The strain that this switcher design effort placed on the engineering staff had not increased its design flexibility. Moreover, the switcher program had cost many hours of valuable engineering time, hours that could have been spent on the design of the system. Finally, in a last minute effort to get the system to the market on time, the company went outside and is currently purchasing its switcher requirements.

Many companies run into these types of problems when attempting to design switchers, because they do not completely understand the huge differences between linears and switchers.

To begin with, switchers must be specified completely differently. Next, they must be tackled by higher levels of engineering. The relatively simple magnetic and thermal problems associated with linear design can easily be handled by junior members of the engineering staff. Switchers, however, in-

Parameter	Linear	Switcher
Incoming Components		
transistor turn on/off	No	Yes
capacitor esr	No	Yes
rectifier leakage	No	Yes
Completed Power Supply		
line/load regulation	Yes	Yes
overvoltage protection	Yes	Yes
current limiting	Yes	Yes
holdup time	No	Yes
power fail alarm	No	Yes
output noise	Yes	Yes
power-on sequencing	No	Yes
dynamic regulation	No	Yes

FIGURE 9.2.
Power supply manufacturing test programs.

volve complex, high-power, nonlinear analog feedback loops that require the efforts of a highly trained specialist. Manufacturing and testing techniques are also significantly different and more complex for switchers (Figure 9.2).

In most major corporations, the responsibility for a make-or-buy decision is commonly assigned to a committee made up of middle-management personnel. The committee usually consists of the program (or project) manager as chairman and representatives from various concerned disciplines, such as product engineering, materials, contracts, manufacturing, industrial engineering, and small business administration. The key individual is the program or project manager.

The factors considered in make-or-buy determinations can generally be classified as nonfinancial or financial in nature. Specific program circumstances will dictate the relative significance of each factor. On occasion, one or more factors may be so significant that it will not be necessary to completely evaluate the remaining factors. The following discussion will focus first on nonfinancial and then on financial aspects of the make-or-buy decision.

NONFINANCIAL FACTORS

Technical considerations determine whether the item or process is feasible from a technical and operational point of view; that is, can the company effectively satisfy its needs and, at the same time, work within the scope of its normal activities? If it can, the decision process can proceed to consider the financial factors.

The analysis of technical capabilities involves comparisons between the company and outside sources. An important factor, especially when considering devices as complex as switchers, is the specific product experience of the engineering staff and of the corporate organization in general. Major switcher users who maintain an engineering capability in the area of power supplies generally plan on a period of 18 months to design and develop a new switcher.

This design and development period can be significantly lengthened by changes in program scope, loss of technical personnel, or the lack of clearly defined objectives and specifications. In addition, specialized test and development facilities must be installed and maintained, and the pool of technical expertise must be constantly updated to reflect the latest developments in a field as dynamic as switchers.

A recently added dimension in the area of technical expertise is complete familiarity with the various regulations that affect switcher design and production. The electromagnetic interference limits imposed by the FCC and the VDE and the safety requirements of the UL, CSA, IEC, and VDE are the most important and complex.

In addition, it is expected that minimum efficiency standards will soon be imposed on the electronics industry, just as they already have been on the auto and electric appliance industries.

Finally, the basic technology of switchers is expected to change significantly in the next few years as bipolar transistors are replaced by power MOSFETS. This change will move the operating frequency of switchers from tens to hundreds of kilohertz and will have a dramatic effect on size and efficiency. For an in-house effort to take advantage of these improvements, a company will require investment of large sums in research and development.

In the long term, this is the most critical of the nonfinancial factors. If a company with an in-house switcher effort fails to keep pace with the technology, its overall system performance will be weakened relative to its competitors. The result will be weakened market position, lost revenues due to lost sales, or even significant loss of market share.

In the short term, however, production capacity and flexibility are the critical nonfinancial factors. Switchers involve high-power, nonlinear analog feedback circuits that require different production techniques than most of the digital systems in which they are employed. The availability of efficient manufacturing, burn-in, and test facilities is a key short-term factor in determining the success of any in-house effort to produce switchers.

FINANCIAL FACTORS

Financial analysis must be completed to determine whether the costs involved with an in-house production effort are reasonable. This is generally performed after determining the feasibility of in-house production from both a technical and manufacturing point of view.

Two types of cost factors must be considered in a complete financial analysis:

1. Direct and indirect costs normally associated with the production process.

2. Subsidiary costs that result from this process but are usually not included with normal direct and indirect costs.

The subsidiary costs of in-house production are difficult to identify and even more difficult to measure accurately. They include, but are not limited to, the following:

1. Experience (learning) curve costs.

2. Absorbing a temporary reduction in work load to retain specialized skills.

3. Equipment and organizational capacities.

4. The ability to efficiently use existing facilities and minimize the investment in new plant and equipment.

5. The placement of work at sources capable of producing follow-up requirements (including warranty and retrofit programs).

The learning curve is the most easily generalized of the subsidiary costs. It relates the cost to produce the next unit to the number of units previously produced. Generally assumed to be log-linear, the typical learning curve states that production costs decline by a constant percentage each time the cumulative production volume doubles (Figure 9.3).

In the case of switching power supplies, the learning curve is a significant cost factor until a minimum of 3000, and more likely 10,000, units have been produced. The actual quantity depends on the complexity of the specific units. Typically, the cost of production declines 3% to 6% each time cumulative volume doubles. For a switcher that has a relatively shallow curve (3%), whose price is $540 after 3000 units have been produced, the initial units would cost between $700 and $747 each.

The subsidiary costs are the hidden costs of a make-or-buy decision. They are difficult to isolate, but their effect on a company's financial bottom line is undesirable. Under the best possible circumstances, the profitability of the primary system will be so great that it will completely overshadow these subsidiary costs. However, if the primary system is marketed in a highly competitive environment, the impact of the subsidiary costs can become significant.

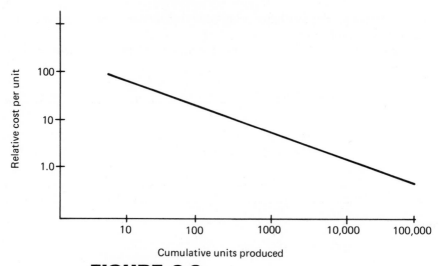

FIGURE 9.3.
Typical power supply production learning
curve.

The financial analysis portion of most make-or-buy decisions ignores the
subsidiary costs and concentrates on the normal direct and indirect costs associated with manufacturing. These normal costs can be broken down into four
primary economic factors:

1. Return on investment
2. Manufacturing costs
3. Inventory costs
4. Warranty costs

Return on investment

The following analysis is typical of those used to determine the financial
feasibility of making instead of buying switching-regulated power supplies, although exactly what constitutes an adequate return on investment (ROI) is a
judgmental decision.

Generally, the level of return is proportional to the risk involved. Investments should be made with regard for the ROI a firm expects to make in its
primary field of activity. This expected return is the *return factor*. It is defined as
the summation of profits over a given period, divided by the initial investment.
To determine the investment cost per unit, the following equation is used.

unit investment cost = discounted capital cost per unit

$$\$/\text{unit} = \frac{I(1 + I)N^N}{(1 + I) - 1} \frac{(P)(R_f)}{X} \frac{I(1 + I)^N}{(1 + I)^N - 1}$$

where P = initial capital investment in design and development (typically $200,000 to $500,000)

R_f = return factor (ranges from 1 to 5, with an average value of 3)

X = switchers used per month

I = interest rate

N = estimated production life (in months) of the primary system

For a product with a 5-year (60-month) production life and a 10% rate of interest, we have

$$\frac{I(1 + I)^N}{(1 + I)^N - 1} = 0.02125$$

and the investment cost per unit becomes

$$\$/\text{unit} = \frac{(P)(R_f)}{X} 0.02125$$

For any given combination of initial investment, return factor, and monthly usage rate, we can determine the investment cost per unit based on the conservative assumptions of a 5-year life and a 10% rate of interest.

From this analysis, it can be seen that the investment cost per unit increases in direct proportion to both the size of the initial investment and the return factor, other things being equal. In a like fashion, the investment cost per unit decreases in direct proportion to the monthly usage rate.

It is difficult to determine, in advance, the size of investment required to design and develop a switching-regulated power supply. However, based on the experience of several OEMs and power supply companies, it is possible to establish a range of typical investment costs. For a switcher, the capital investment can range from $200,000 to $500,000 or more. This amount would be evenly split between design engineering and manufacturing development (Figure 9.4).

The design engineering portion of the investment includes the specialized testing and analysis equipment required for a switcher project but not required for projects involving digital systems. The manufacturing development costs include tooling, pilot runs, specialized burn-in and test fixtures, and so on. The facilities required for the efficient design and production of high-power, nonlinear analog feedback systems (switchers) are significantly different from those required when working with digital systems.

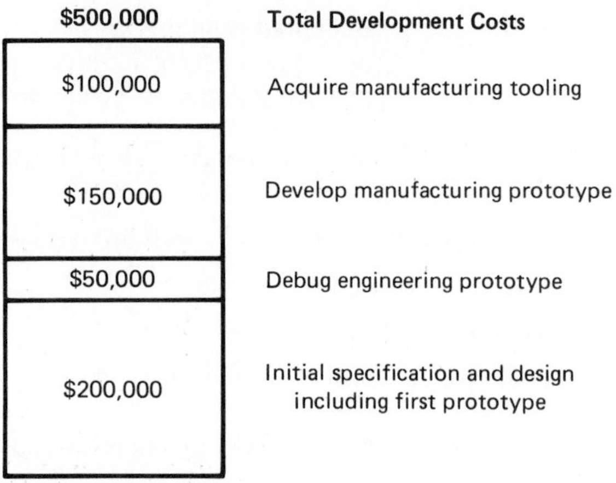

$500,000	Total Development Costs
$100,000	Acquire manufacturing tooling
$150,000	Develop manufacturing prototype
$50,000	Debug engineering prototype
$200,000	Initial specification and design including first prototype

FIGURE 9.4.
Typical development costs for a switching regulated power supply.

Manufacturing costs

Manufacturing cost is composed of three primary element' labor, material, and overhead. A typical power supply has a manufacturing cost of at least 60% of the selling price. This figure can vary with the size (wattage) of the power supply and from one company to another. Manufacturing overhead rates for power supply companies average about 150%. Material costs also vary from company to company but average around 33%.

Given the preceeding averages for manufacturing costs, overhead rates, and material costs, it is possible to determine the breakdown of labor, material, and overhead contained in the manufacturing cost. If the total manufacturing cost is 60% of the selling price and material represents 33% of that same price,

Overhead Rate	150%	200%	250%	300%	350%
Labor	10.8%	10,8%	10.8%	10.8%	10.8%
Overhead	16.2%	21.6%	27.0%	32.4%	37.8%
Material	33.0%	33.0%	33.0%	33.0%	33.0%
Manufacturing cost	60.0%	64.5%	70.8%	76.2%	81.6%

FIGURE 9.5.
Manufacturing cost vs. overhead.

then 27% remains for labor and overhead expenses. With an overhead rate of 150%, 10.8% must go for labor while 16.2% goes for overhead.

This distribution of the total manufacturing cost (60% of the selling price) into 10.8% for labor, 33% for material, and 16.2% for overhead provides a basis from which the effect of different overhead rates on manufacturing cost can be examined. If labor and material costs are assumed to be relatively fixed, then manufacturing costs will vary from 60% to 81.6% as overhead varies from 150% to 350% (Figure 9.5).

Inventory costs

Inventory costs can be broken down into two factors: incremental inventory needed to support the make decision, and the costs of holding inven-

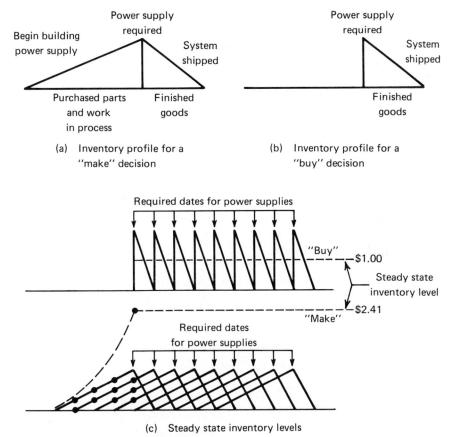

(a) Inventory profile for a "make" decision

(b) Inventory profile for a "buy" decision

(c) Steady state inventory levels

FIGURE 9.6.
Inventory costs of a make-buy decision.

tory. Incremental inventory consists of the additional purchased parts and work in the process required to support a switcher production. The inventory of finished power supplies must be held independently of the decision to make or buy.

In the case of a make decision, inventories must be built up in advance of actual power supply need. If the assumption is made that finished goods account for 20% of the total inventory, and if the manufacturing cost is 76.2% (overhead is 300%, typical for an OEM), it can be shown that for every $1 of purchase price an additional investment of about $1.41 in inventory is needed in the form of purchased parts and work in process (Figure 9.6).

These inventories of parts, work in process, and finished goods must be financed. Inventory carrying costs can be broken down into two components, the opportunity cost and the direct cost. The opportunity cost equals the inventory turnover rate times the average net profit per shipment. For a turnover rate of 3 and an expected average profit of 10%, the opportunity cost of holding additional inventory is 30% per year. Actual opportunity costs can vary significantly, but generally fall in the range of 20% to 50% per year.

The direct costs of carrying inventory consist of storage costs, handling, taxes, insurance, and obsolescence. These factors usually total 20% to 25% per year. The total inventory holding cost is the sum of the opportunity cost and direct cost. Therefore, the combined inventory cost can range anywhere from 40% to 75% per year, with 60% per annum (5% per month) being a reasonable average value.

Warranty costs

Warranty costs are minor compared to the other costs discussed, but they must be added to complete the analysis. They generally run 3% to 5% of the purchase price of a switcher, based on a warranty period of 2 years.

Making a decision

To summarize the financial analysis, consider the make-or-buy decision for a company that requires 1000 four-output, 375-W power supplies per year. The known factors are as follows:

1. Design and development investment: $200,000
2. Return on investment factor: 3
3. Manufacturing overhead: 300%
4. Finished goods ratio: 20%
5. Annual interest rate: 10%
6. Purchase Price (1000 pieces): $540
7. End product production life: 5 years

8. Length of switcher manufacturing cycle: 8 weeks

$$\text{Investment cost/unit} = \frac{(\$200,000)(3)}{83} \ 0.02125 = \$153.61$$

$$\text{Manufacturing cost} = (540)(0.762) = 411.48$$
$$\text{Inventory cost} = (540)(1.41)(0.05) = 38.07$$
$$\text{Warranty cost} = (540)(0.03) = \underline{\$16.20}$$
$$\text{Total cost to build} = \$619.36/\text{unit}$$

In this example, it is clearly more cost effective to buy switchers. As production volume increases, however, the unit cost decreases until, at some point, it is less expensive to make switchers. This economic transition between make and buy occurs at a usage rate of over 2000 units per year.

It is important to remember the major assumptions on which the economic transition is based: a conservative investment estimate of only $200,000 (it could easily go as high as $500,000); an interest rate of only 10% (real interest rates will probably be higher); a product life of 5 years (this is an area of significant risk in today's rapidly changing marketplace); and three inventory turns per year (not always accomplished). If any of these assumptions prove to be too optimistic, the economic transition volume will increase beyond 2000 units per year. In a real-world situation, it is not unusual for the transition to occur at a volume in excess of 5000 to 10,000 units per year.

In addition to the cost savings of $79.36 in the previous example, the lead time to buy OEM quantities of switchers from power supply manufacturers ranges from 2 months for a standard model to 5 months for a custom design. To design a similar unit from scratch would require 12 to 24 months of intense effort. Even if production volumes are at or above the economic transition level, the problem of lead time will remain.

SUMMARY

The purpose of this discussion has been to point out potential pitfalls and to try and eliminate the blind spots that face a company contemplating the in-house production of switchers.

The make-or-buy determination is one of the most common of all business decisions and one of the most complex. In the case of switching regulated power supplies, it requires a very close examination of specific technical capabilities within the organization. Only after there is reasonable assurance of adequate technical capability should the financial aspects be considered. In the end, however, the financial aspects of the make-or-buy determination are at least as important as technical capability in arriving at a sound business decision.

TRENDS IN TECHNOLOGY

Power supply technology, like almost all areas of electronics, is not fixed or unchanging. In fact, important evolutionary changes are taking place that will change power supplies in the next 5 years as much as they have changed in the preceding 10 years.

Switching supplies have been intriguing to systems engineers for many years because of their higher efficiency and more compact, cooler-running packages. Unfortunately, the trade-offs in terms of reliability and cost were too great to allow switchers to compete effectively with linears in all but a few very specialized applications.

The result is that two distinct segments make up the majority of all power supply usage. Linears are by far the largest portion of the market and currently account for around 60% to 70% of all power supply usage. Switchers make up most of the balance, with ferros and miscellaneous types making up only about 5%.

Linear power supply techniques are well known and are a mature technology. The usage of linears is increasing only about 8% to 10% per year. Switch-

147

ers, on the other hand, incorporate a still developing technology and switcher usage is growing at a 25% per year rate. The significant difference in growth rates is causing basic changes to occur in power supply usage.

These changes have important implications for power supply manufacturers and specifiers. In the short term, the increasing use of switchers will mean a larger and larger potential market for manufacturers of power supplies, especially switchers. The reason for this market growth is that OEM power supply users are less likely to manufacture switchers in-house, and the growing use of switchers will have to be satisified by turning to outside vendors. In the longer term (5 to 10 years), however, switcher technology will become simpler to implement owing to increasing use of better ICs and to a generally better understanding of switchers. As a result, the current rapid growth in the noncaptive sector of the power supply market is not expected to continue beyond the near term.

TODAY'S TRANSITION

Today, most power supplies employ the linear or series-pass regulation technique. Linears are inherently low-noise devices and have much faster transient response times than switchers, but they also have a number of characteristics that are becoming larger and larger drawbacks. When energy and raw material costs were relatively low, it did not matter that a typical linear dissipates 60% of its input energy as heat; nor was the large amount of copper and steel used to manufacture linears considered a serious disadvantage.

Now that neither energy nor raw materials are as cheap as they once were, the inefficiencies of linears are a significant consideration when choosing a power supply regulation technique. In response, the industry is shifting to a newer and more efficient form of regulation called pulse-width modulated (switching) regulation.

Switchers are relatively new only in industrial and commercial applications; they have been used by the military for well over 20 years. It took the development of more reasonably priced switching transistors and support components to make switchers commercially viable. In fact, the first commercial switchers incorporated transistors developed for television and automotive applications. To achieve reasonable performance, these transistors had to be individually matched for switching characteristics, resulting in an extremely high and costly rejection rate.

The second major problem with early commercial switchers was the lack of support components. Where a handful of ICs is used today, early switchers had a mass of discrete components. Switchers are more complex than linears, and the lack of support components for early design caused switcher prices to be quite high compared with the simpler linears. Initially, switchers were competitive with linears only in the high-wattage (250+ watts) end of the market.

Ongoing improvements in switcher components have made today's power supply market quite dynamic. As switcher sales volume increased, demand for switching transistors grew to the point where semiconductor manufacturers began to design and produce devices specifically for switching power supplies. In addition to decreasing switcher cost, the new, specially designed components also increased switcher reliability.

Increased sales volume has prompted the development of ICs specifically for use in switchers. These ICs significantly reduce the cost of designing and manufacturing switchers. Furthermore, the use of components developed specifically for switchers and the increasing use of ICs to reduce component count have improved switcher reliability to the point that it is now comparable to that of linears.

This continuing adoption of switching regulation is having significant impact on power supply users and manufacturers. For power supply users, the make-or-buy analysis for power supplies comes up "buy" more and more often as switchers are used in place of the simpler linears. This is causing the OEM market for switchers to grow much more rapidly than the market for linears. This difference in growth rates is having a major impact on power supply manufacturers.

An important result of the trend away from linears and toward switchers is that industry leadership will shift as the basic technology changes. Today's major linear power supply manufacturers could easily be replaced by a new group of switcher manufacturers. This is not an unknown phenomenon. It has happened many times in the electronics industry.

Examples include the displacement of RCA, Sylvania, and General Electric as suppliers of vacuum tubes by Texas Instruments, National Semiconductor, and Motorola as suppliers of semiconductors; and the replacement of Friden, Monroe, and Victor as suppliers of electromechanical calculators by Texas Instruments, Hewlett-Packard, and Casio as suppliers of electronic calculators.

Leadership changes as technology changes. Companies that hold a large share of a market for current technology products, such as linears, have major technical and financial investments that cannot be adapted quickly or easily to changing technology. Major suppliers must precisely time their change to a new technology, such as switching regulation. Changing before the new technology has been accepted by a wide enough segment of the market or after competitive firms have become entrenched in the new marketplace can cost companies a substantial portion of their market share.

Switcher technology itself is expected to improve significantly in the future with the replacement of bipolar transistors by power metal oxide semiconductor field effect transistors (MOSFETs). This change will be gradual, beginning with low-power switchers and slowly being incorporated into higher-wattage units. To achieve long-term survival as a major power supply manufacturer, a company must not only adopt switcher technology, it must also make the transition from bipolar to MOSFET switchers, with the sizable research and development investment that entails. Again, those manufacturers who cannot meet the challenge will lose market share or worse.

TOMORROW'S SWITCHERS

Three years ago, almost all commercial switchers operated at 20 to 25 kHz. Today there are high-power units operating at 50 kHz and some low-power units at over 200 kHz. Tomorrow, switcher frequencies of over 200 kHz will be the rule rather than the exception. These higher frequencies will decrease output ripple and improve the transient response times of switchers, narrowing the advantage that linears have in those areas. In addition, high operating frequencies will enhance switcher advantages such as small size and light weight and will lower switcher prices relative to linears.

Power MOSFETs will offer continuing improvements in switching supplies because of their higher switching speeds, which are due to their majority-carrier nature. With high-frequency operation, smaller transformers, inductors, and capacitors will reduce power supply size and will lower cost. In contrast to bipolar transistors, which are current-controlled devices, MOSFETs are voltage controlled and require very little drive current. Switcher drive circuits, which now use discrete components, will switch to ICs, just as switcher control circuitry did 5 years ago. The simplified drive circuitry needed by MOSFETs will increase switcher reliability, while at the same time lowering costs.

The negative temperature coefficient of MOSFETs makes them even more attractive for use in switchers. The most common failure mode in today's switchers is second breakdown of voltage in power transistors caused by uneven conduction. Since MOSFETs do not develop localized hot spots, they are not susceptible to second breakdown. Properly designed switchers incorporating MOSFETs should be even more reliable than today's highly reliable bipolar-based switchers.

There are still some problems to be overcome prior to universal adoption of MOSFETs by switcher manufacturers. Power MOSFETs suffer from the disadvantage of high on-state resistance as compared to bipolar transistors. However, power MOSFET manufacturers are overcoming this problem. A few years ago, typical on-state resistance for MOSFETs was 4 to 5 Ω; today it is well below 0.1 Ω and is closing on that of bipolar transistors.

A more important problem is the relatively low reverse-voltage tolerance of power MOSFETs. This problem is yielding much more slowly and represents a major stumbling block to the widespread use of MOSFETs in switchers. Large reverse-voltage spikes appear in the high-voltage switching section of off-line switchers. Power MOSFETs have very fast switching times, making these inductance-caused spikes even worse. Power supply designers will have to pay attention to the placement and design of *snubbers* in circuits containing MOSFETs.

The transition from bipolar to MOSFET switchers will resemble the replacement of junction rectifier diodes by Schottky devices. Before MOSFETs can replace bipolar transistors in the majority of switching power supply designs, the MOSFETs themselves must be improved to reduce their shortcomings in high-voltage, high-frequency switching applications. Also, the power supply industry must develop new design techniques to take full advantage of the performance improvements promised by MOSFETs.

Their somewhat higher on-state resistance and relative sensitivity to reverse-voltage spikes has, so far, limited the use of MOSFETs to low-power applications, which put less demands on switching components. As MOSFET technology matures and switcher manufacturers continue to gain design experience with them, MOSFETs will replace bipolar transistors as the standard switching component used in switchers.

The first generation of switching regulator ICs appeared in 1976 and greatly simplified the design and production of switching regulated supplies. Today, IC manufacturers offer much improved versions with more on-chip protection circuits, higher drive current (in some cases capable of straight-driving the switching transistors), single- or double-ended outputs, variable-frequency capability, and greater reference-voltage accuracy.

Families of switcher ICs are now available that include a variety of support chips to work with the main pulse-width modulation control chip. Specialized driver ICs provide higher output currents. Voltage-reference chips provide very accurate ($\pm 0.25\%$) references. Separate ICs are available for protection functions, which include current sense, overvoltage protection, undervoltage sense, current limit, power fail, and fault activation delay or logic-level-fault indication outputs.

As the operating frequency of switchers increases, power transformer leakage inductance will become a problem that can reduce overall efficiency. To maintain high efficiencies, power transformer manufacturers must refine their products to reduce leakage inductance. Higher operating frequencies will also mandate the need for small transformers with fewer windings, increasing the problems of transformer design.

The shrinkage of the size and number of windings will make it essential to design transformers with better primary to secondary coupling. Improved winding geometries will be required to achieve the necessary primary to secondary coupling. The industry will continue to refine ferrite cores, which operate efficiently at high frequencies and high flux densities. Small transformers will be possible at higher-frequency operation only if efficiency remains high and operating losses (heat) are minimized.

For a given level of ripple, a switcher's output inductor size decreases as the operating frequency increases. The output filter inductor is optimized at the primary ripple frequency, and its inductance strikes a balance between low output ripple and fast dynamic response. Since a switcher must wait for the next half-cycle of switching frequency before starting to regulate for changes in output voltage, it may take 3 to 4 cycles before a step-load change is regulated. Therefore, for a given level of ripple, the dynamic response is proportional to operating frequency. If a 20-kHz switcher has a dynamic response time of 200 μs, a 200-kHz switcher will respond in 20 μs at equal output ripple levels. By operating at higher frequencies, switchers will reduce the advantage that linears enjoy in lower output ripple and faster dynamic response.

Specialized capacitors are also available for use in switchers. Manufacturers have developed balance capacitors for the power switching section of half- and full-bridge switchers. Used in series with the primary winding of the power transformer, these capacitors are critically important in *balancing* the volt-seconds across the transformer in both directions. Since it is not economically practical to precisely match power transistor switching characteristics used in each power supply produced, the capacitor performs a balancing function. Unless the volt-seconds are properly balanced, the power transformer core can saturate in one direction. If that happens, the transistor(s) switching on in that direction will see a short-circuit condition and can go into second-breakdown mode. The balance capacitor helps to prevent second breakdown from occurring.

The development of low equivalent series resistance capacitors has also been important in the growing success of switchers. Although not specifically designed solely for switchers, low ESR capacitors in the input and output filter sections of off-line switchers help to maintain high operating efficiency. The size of the input capacitors is related to energy storage and holdup time, not operating frequency. The output filter capacitors used at higher frequencies will be quite different.

Since the output filter requires less storage capacity between ripple peaks at higher operating frequencies, high-frequency switchers can use smaller capacitors and no longer need large, relatively expensive electrolytic filter capacitors. Polypropylene film or similar capacitors that are better suited for high-frequency operation will replace today's electrolytics. Polypropylene film capacitors are significantly smaller and less expensive than electrolytics.

EXTERNAL FACTORS

External factors are influencing and accelerating the move to switchers. The quality of the electric power delivered on the major power distribution networks is only fair. It is quite common for ac line voltage to skip a cycle so that a 16-ms power dropout occurs. Dropouts may be caused by power distribution network switching or a natural phenomenon like lightning.

Most digital systems cannot tolerate a loss of power for 16 ms without deteriorated performance, and linear power supplies cannot provide sufficient holdup in the case of line voltage dropouts. Switchers, however, typically provide 20 ms of holdup, enough to protect digital electronics from line voltage dropouts. Switchers also protect systems from brownout conditions. The input voltage range for a typical linear is $\pm 8\%$; the corresponding range for a switcher is $+10\%$ to -20%.

It is generally accepted that rolling brownouts and dropouts of ac line power will continue to be common occurrences in the future. These factors, in addition to the increasing importance of digital systems to business and society in general, will contribute to accelerating the rate at which switchers displace linears as the dominant power supply technology.

Changes in the scope of government regulations are also affecting the transition. The Federal Communications Commission (FCC) has adopted limits on the amount of electromagnetic interference that can emanate from electronic systems. Most current-technology switchers must have added filtering to meet the FCC requirements. Extra filtering means extra cost. Linears are inherently EMI-free devices and require little or no filtering. The added filtering requirements will increase the cost of switchers relative to linears and possibly have a negative impact on the growing use of switchers. As switchers move to higher operating frequencies, however, the filtering problem (and the associated costs) will decrease significantly.

Another important area of government regulations is international safety standards. Because of the relatively large number of components in the primary circuit, switchers will have a more difficult time meeting these safety standards than will linears. Much of the impact occurs in the design stage, stretching out the design cycle for switchers, since the safety standards tend to significantly complicate the design process. This factor will tend to increase the demand for noncaptive switcher production since OEMs will be reluctant, in some cases, to spend the additional design time and expense.

Linears, which are simpler and have fewer primary components, will be less affected in terms of design impact. The impact of the international safety standards will be much the same for both linears and switchers at the production stage. The net impact of this factor will be to strengthen the noncaptive switcher manufacturers relative to in-house switcher efforts, but it should have no significant impact on the transition from linears to switchers.

A final area of government regulation that may have an effect is the minimum allowable efficiency standards for electronic systems and equipment. Efficiency standards have already been imposed on the auto industry and are pending for household appliances. If another oil embargo or similar energy crisis occurs in the near future, electronic equipment could become regulated for minimum allowable efficiencies. Any reasonable requirement, such as 50% or greater efficiency, could mandate the use of switchers instead of linears.

The speed and direction of technical change in the entire electronics industry will affect the transition. All manufacturers will be forced to work harder than ever just to keep up with changes in their own fields of specialization. Electronic OEMs now producing power supplies (usually linears) in-house as a sideline will no longer be able to keep abreast of current technology in both their primary field and in the area of power supplies. The result will be the growth of the noncaptive (switcher) segment of the power supply market.

The rising costs of energy and raw materials will tend to accelerate the shift to switchers. Although the initial purchase price is important, power supply efficiency can have a greater impact on the overall cost of ownership. The efficiency of a power supply is often oversimplified and misunderstood. It should be considered in both electrical and material terms.

The electrical efficiency of a power supply is generally defined as the ratio of output power to input power. By this measure, the efficiency of a typical linear is 40% and that of a switcher, 70%. The true magnitude of the difference between 40% and 70% efficiency may not be obvious. Consider two typical 500-W power supplies, a linear and a switcher. The linear requires 1250 W of input power, the switcher, 714 W. Energy consumption of these units differs significantly in that the switcher dissipates only 214 W of input power, while the linear dissipates 750 W, more than three times as much. In terms of today's energy costs, that is an expensive difference over the entire operating life of the systems in which the power supplies are installed.

Another important consideration is the need to exhaust the heat dissipated by each power supply. Most modern electronic systems operate in environmentally controlled surroundings. A linear places a much greater load on air-conditioning systems than a cooler-running switcher. As a result, although the difference between 40% and 70% efficiency is less than a factor of 2, it translates into a fourfold difference in dissipated energy. That fourfold difference is becoming increasingly important as the cost of energy continues to rise.

The second aspect of efficiency is material efficiency. Switcher prices, now competitive with linears at levels above 50 W, will continue to fall as the technology matures. Linears rely on large 60-Hz power transformers, which consume large quantities of steel and copper. Therefore, the prices of linears will continue to rise with the increasing prices of raw materials. In addition to being costly, large 60-Hz transformers are heavy. A linear typically delivers 5 W/lb; a switcher, with a smaller 20-Hz transformer, delivers 20 W/lb. Physical volume is another manifestation of a power supply's material efficiency. Linears typically achieve 0.33 W/in.^{3}, switchers, 1.0 W/in.^{3}. A switcher will weigh one-fourth as much and occupy one-third the volume of a comparable linear. These are increasingly important differences in the rapidly shrinking world of electronics.

IMPLICATIONS FOR OEMS

During the transition from linears to switchers, the latter will be in short supply. The major switcher manufacturers have not yet grown large enough to keep up with demand, and the leading linear companies have yet to complete the transition. This situation has created the opportunity for numerous small switcher companies, which have sprung up to fill the gap between the supply of and the demand for switchers. As the transition proceeds toward completion, a major shakeup of the power supply industry is likely to occur. Small switcher manufacturers may be severely tested. Some large producers of linears will successfully make the transition; others will not be as fortunate.

The transition from bipolar to MOSFET technology within the switcher industry could intensify the coming shakeup and affect power supply users, as well as producers. Users who purchase their requirements from outside vendors should adapt their power supply specification and inspection procedures to switcher technology and review their list of qualified vendors with an eye to switcher capability.

OEMs who are also in-house producers of power supplies will also face ever increasing difficulties in manufacturing state-of-the-art power supplies. Linears can be designed in a matter of weeks, but a switcher design requires many months. This fact will encourage many firms that are now producers to turn to outside suppliers to overcome the problems of efficient power supply production. Moreover, changes in government regulations and other external factors will compound the problems of firms that pursue in-house production of switchers.

The performance of switching regulated power supplies will make a major advance as a result of the new MOSFET power semiconductors that will replace today's bipolar devices. High-frequency operation (over 200 Hz) will result in a significant decrease in the sizes of many passive components. Smaller power transformers, filter capacitors, and inductors will contribute to lower costs, as well as improve switcher compactness.

Operating parameters, such as transient response time and output ripple, will also improve as a result of higher operating frequencies. These are two of the major advantages that linears currently enjoy over switchers. While switchers will probably not equal linears in either transient response or output ripple in the near future, the gap will be made much narrower.

Typical peak-to-peak output ripple for low-voltage switchers is 50 mV. As semiconductor companies develop new generations of MOS devices for use in digital systems, the industry-standard supply voltage is expected to drop from today's 5 V to between 2 and 3 V. At outputs of 2 to 3 V, 50 mV of ripple is usually unacceptable. The new high-frequency switchers currently under development will allow the reduction of output ripple and, at the same time, improve dynamic response characteristics.

High-frequency operation will also lessen the problem of switcher-produced EMI. Today, extensive filtering is generally necessary to bring switchers into compliance with the requirements of the FCC and VDE. Most of the filtering difficulties arise owing to the lower-level harmonics associated with operating at 20 kHz. When operating at 200+ kHz, those problems will be significantly reduced, and so will the size and complexity of the needed EMI filter.

Finally, the increasing use of switchers is forcing OEMs to change their power supply burn-in and testing procedures. Solid-state programmable burn-in loads and automated test equipment are needed when working with switching power supplies. Most linears are relatively simple devices and can be adequately checked out with a resistive load and a manual test bench. The complex and dynamic characteristics of switchers, such as transient response, timing power-fail signals, output voltage sequencing, and the like, are much more difficult to efficiently test in high volumes without automation. A side effect of the growing use of sophisticated switchers is a relatively new and growing market for automated power supply testers and programmable burn-in loads.

SUMMARY

The future will see switchers and linears coexisting, but with switchers displacing linears as the dominant power supply technology. Switchers will offer greater efficiency, more compactness, and improved ripple and dynamic response, and will cost less than linears. Linears will still have better dynamic response and lower output ripple than switchers and will be used, primarily, in special applications (e.g., low-level analog) that require minimal levels of ripple or very fast dynamic response speeds.

OEM users of power supplies are having to adapt to the transition from linears to switchers in a number of ways. The process of power supply selection, specification, qualification, and application is becoming increasingly complex. Switchers incorporate a higher level of technology than linears and increasingly offer the system designer many advantages over linears.

Not only are improvements in power supplies affecting the rest of the electronics industry, but changes in technology in other parts of the electronics industry will affect the power supply market. The most important will be the continuing development of microprocessor-based systems, which will cause the market for high-wattage power supplies to become a smaller portion of the overall market. The net result of all these changes will be that two kinds of power supply manufacturers will most likely survive in the long term: small but technically strong companies that produce a low volume of high-power and custom switchers, and a few large companies that produce a very high volume of low-power switchers for the microprocessor-based systems currently under development.

GLOSSARY

Ambient temperature: The temperature of the environment surrounding a power supply, generally assumed to be room temperature.

Bandwidth: Based on the assumption that a power supply can be modeled as an amplifier, the bandwidth is that frequency at which the voltage gain has fallen off by 3 dB. Bandwidth is an important determinant of transient response and output impedance.

Bleeder resistor: A resistor usually connected across a filter circuit to discharge capacitors when the unit is turned off.

Breakdown voltage: See *Isolation.*

Bridge: (1) Rectifier circuit incorporating four diodes (full-bridge) or two diodes (half-bridge) (Chapter 2). (2) Converter or chopper section of switching power supplies incorporating four transistors (full-bridge) or two transistors (half-bridge) (Chapter 3).

Brownout: Condition during peak usage periods when electric utilities reduce their nominal line voltage 10% to 15%.

Brownout protection: The ability of a power supply to continue operating within specification through the duration of a brownout.

Brute-force supply: The most basic form of power supply, which delivers an unregulated dc output voltage (Chapter 2).

Burn in: The period directly following the very first turn on of a given power supply. It is characterized by a relatively high and declining failure rate (Chapter 5).

Bus: (1) The system of conductors (wire, cable, copper bars, etc.) used to transport power from the power supply to the load. (2) A communications structure used to control various instruments and subsystems (e.g., IEEE-488 bus).

Carry-over: See *holdup time.*

Centering: The variation of an output from its nominal voltage caused by design limitations and manufacturing variations. Centering is expressed as the percentage ratio of the voltage deviation to nominal output voltage (Chapters 4 and 6).

Chopper: See *Inverter.*

Common-mode noise: That component of noise common to the output and return lines with respect to an electrically fixed point, usually chassis ground (Chapters 4 and 6).

Constant current: A power supply that regulates current level regardless of changes in load resistance.

Constant current limiting circuit: Current-limiting circuit that holds output current at some maximum value whenever an overload of any magnitude is experienced (Chapter 7).

Constant voltage: A power supply that regulates voltage level regardless of changes in load resistance.

Convection: The transference of thermal energy in a gas or liquid by currents resulting from unequal temperatures (Chapter 5).

Converter: (1) A device that delivers dc power when energized by a dc source. (2) Sections of a switching power supply that perform the actual power conversion and final rectification (Chapter 3).

Cooling: Removal of heat, which, in a power supply, is generated by transformation, rectification, regulation, and filtering. It can be accomplished using radiation, convection, forced air, or liquid means (Chapter 4).

Cross regulation: In a multiple-output power supply, the load variation of one output can cause a voltage change in other outputs. This voltage change divided by its nominal value is the cross regulation (Chapter 5).

Crowbar: A type of overvoltage protection in which an SCR is placed directly across the output terminals of a power supply (Chapter 7).

CSA (Canadian Standards Association): An independent Canadian organization testing for public safety, similar to the function of Underwriters' Laboratories in the United States (Chapter 8).

Current limiting circuit: A bounding circuit designed to prevent overload of a constant-voltage power supply. It can take the form of constant, foldback or cycle-by-cycle current limiting (Chapter 7).

Cycle-by-cycle current limiting circuit: Current-limiting circuit that immediately reduces output current to some minimum level whenever an overload of any magnitude is experienced (Chapter 7).

Derating: A reduction of some operating parameter to compensate for a change in one or more other parameters. In power supplies, the output power rating is generally reduced at elevated temperatures (Chapter 4).

Dielectric withstand voltage: See *Isolation.*

Differential-mode noise: That component of noise measured with respect to output return; it does not include common-mode noise (Chapters 4 and 6).

Drift: See *Stability.*

Dynamic load: A load that rapidly changes from one level to another. To be properly specified, both the total change and the rate of change must be stated.

Efficiency: The ratio of output power to input power. It is generally measured at full-load and nominal line conditions. In multiple-output switching power supplies, efficiency can be a function of total output power and its division among the outputs (Chapters 2 through 6).

EMI (electromagnetic interference): Also called radio-frequency interference (RFI), EMI is unwanted high-frequency energy caused by the switching transistors, output rectifiers, and zener diodes in switching power supplies. EMI can be conducted through the input or output lines or radiated through space (Chapter 8).

ESR (equivalent series resistor): The amount of resistance in series with an ideal (lossless) capacitor, which duplicates the performance of a real capacitor. In general, the lower the ESR, the higher the quality of the capacitor and the more effective it is as a filtering device. ESR is a prime determinant of ripple in switching power supplies (Chapter 6).

Faraday shield: An electrostatic shield wound on a transformer, designed to reduce interwinding capacitance. The result is less common- and differential-mode noise at the output of the power supply.

Federal Communications Commission (FCC): United States federal regulating body whose new EMI limitations are affecting the design and production of digital electronics systems and their related subassemblies, such as power supplies (Chapter 8).

Ferroresonance: The principle used in a simple open-loop (nonfeedback) voltage-stabilizing power supply (Chapter 2).

Filter: A frequency-sensitive network that attenuates unwanted noise and ripple components of a rectified output (Chapter 2).

Flyback converter: Switching power supply configuration using a single transistor and a flyback diode (Chapter 3).

Foldback current limiting circuit: Current-limiting circuit that gradually decreases the output current under overload conditions until some minimum current level is reached under a direct short circuit (Chapter 7).

Forward converter: Switching power supply configuration using a single transistor (Chapter 3).

Frequency changer: Power-conversion equipment that transforms ac electric power from one frequency to another without affecting its other characteristics (Chapter 2).

Full-bridge converter: Four-transistor switching power supply configuration used to handle high power levels (Chapter 3).

Full-wave rectifier: Rectifier circuit that rectifies both halves of an ac wave (Chapter 2).

Ground loop: A feedback problem caused by two or more circuits sharing a common electrical line, usually a common ground line. Voltage gradients in this line caused by one circuit may be capacitively, inductively, or resistively coupled into the other circuits via the common line. With power supplies, this problem can be reduced using single-point grounding (Chapter 7).

Half-bridge converters: Two-transistor switching power supply configuration used in medium-power applications (Chapter 3).

Half-wave rectifier: Single-diode rectifier circuit that rectifies only one-half the input ac wave (Chapter 2).

Head room: In a linear regulator, the head room is the difference between the secondary voltage supplied by the power transformer at nominal input voltage and the regulated output voltage. Head room is necessary to ensure proper regulation under full load and low input voltage operation (Chapter 2).

Heat sink: Device used to conduct away and disperse the heat generated by electronic components (Chapter 5).

Hi-pot (high potential voltage): Ability of a power supply to withstand a high voltage potential placed either from the input terminals to ground, from any of the output terminals to ground, or between any pair of input and output terminals. This specification is important for safety reasons and is partially dependent on the mechanical design of the power supply (Chapter 5).

Holdup time: The total time any output will remain within its regulation band after the input line voltage has been turned off. Typically measured at full load and nominal line conditions (Chapters 4 and 6).

Hybrid supplies: A power supply that combines two or more different regulation techniques, such as ferroresonant and linears or switching and linear (Chapter 3).

Hybrid thermal design: A power supply that uses a combination of convection and forced-air cooling (Chapter 5).

Inhibit: The ability to electrically turn off the output of a power supply from a remote location (Chapters 4 and 6).

Input surge current: See *Inrush current.*

Input voltage range: The range of line voltages for which the power supply meets its specifications (Chapters 4 and 6).

Inrush current: A high surge of input current that occurs in switchers and occasionally in linears upon initial turn on (Chapter 4).

Instantaneous current limiting circuit: See *Cycle-by-cycle current limiting circuit.*

Interaction: Total static regulation of a power supply when line and load changes occur simultaneously (Chapters 4 and 6).

International Commission on Rules for the Approval of Electrical Equipment (CEE): A regional, European safety agency in which the United States participates only as an observer (Chapter 8).

International Electrotechnical Commission (IEC): An international safety agency headquartered in Geneva, Switzerland (Chapter 8).

Inverter: (1) A device that delivers ac power when energized from a source of dc power. Inverters may be frequency, amplitude, or pulse-width modulated to vary output voltage. (2) The chopper section of a switching power supply (Chapter 3).

Isolation: The degree of electrical separation between two points. It can be expressed in terms of voltage (breakdown), current (galvanic), or resistance and/or capacitance (impedance). In power supplies, the input-to-output isolation is important to maximize (Chapters 6 and 8).

Leakage current: Current flowing between the output buses and chassis ground due to imperfections in electronic components and designs. It must be tightly controlled to satisfy safety regulations such as UL and VDE (Chapters 6 and 8).

Line-frequency regulation: The variation of an output voltage caused by a change in line input frequency, with all other factors held constant. This effect is negligible in switching and linear power supplies, but it is a critical specification of ferroresonant power supplies.

Line regulation: The variation of an output voltage due to a change in the input voltage, with all other factors held constant. Line regulation is expressed as the maximum percentage change in output voltage as the input voltage is varied over its specified range (Chapters 4 and 6).

Line regulator: Power-conversion equipment that changes the degree of regulation, filters noise, and/or changes the voltage of incoming ac power (Chapter 2).

Linear regulator: A common voltage-stabilization technique in which the control device (usually a transistor) is placed in series or parallel with the power source to regulate the voltage across the load. The term "linear" is used because the voltage drop across the control device is varied continuously to dissipate unused power (Chapters 2 and 3).

Load: For voltage-regulated power supplies, the load is the output current.

Load regulation: Variation of the output voltage due to a change in the output's load from no load to full load, with all other factors held constant. It is expressed as a percent of the nominal dc output voltage (Chapters 4 and 6).

Logic enable: The ability to turn a power supply on and off with a TTL signal. A logic low generally turns the supply off; a logic high turns it on. See also *Logic inhibit* (Chapter 4).

Logic high: A voltage of greater than 2.3 V with a maximum of 5.5 V. Also known as logic 1.

Logic inhibit: The ability to turn a power supply off and on with TTL signals. A logic low allows the power supply to operate. A logic high turns off the power supply. See also *Logic enable* (Chapter 4).

Logic low: A voltage of less than 0.8 V. Also known as a logic 0.

Margining: The ability to adjust (usually with a switch) the output manually, usually to within ±5% of nominal. This capability is used in system testing (Chapters 4 and 6).

Master: The unit in a master–slave system of interconnection that exercises control over the outputs of one or more slave units. Such a system is a common technique used to ensure load sharing of parallel operating power supplies (Chapter 7).

Modular: A physically descriptive term used to describe a power supply made up of a number of separate subsections, such as an input module, power module, or filter module. Modular construction tends to lower the MTTR.

MTBF (mean time between failures): A measure of reliability. The reliability interval calculated in accordance with the procedures of MIL-HDBK 217 (Chapter 5).

MTTR (mean time to repair): The average time required to repair a power supply. It is a result of both electrical and mechanical design factors.

Multiple output supply: A power supply that delivers two or more different output voltages (Chapter 3).

Noise: Noise is the aperiodic, random component of undesired deviations in output voltage. Usually specified in combination with ripple. See *PARD* and *Ripple* (Chapters 4 and 6).

Nominal output voltage: The intended, ideal voltage of any given output.

Off-line switcher: A circuit configuration commonly used in PWM switchers in which the input rectifier and filter sections sit directly across the ac input line (Chapter 3).

Open-frame construction: A construction technique common to OEM power supplies where the supply is not provided with an enclosure. It can be either a simple printed circuit board or a circuit board mounted on a metal chassis without a cover.

Operating temperature: The range of temperatures within which a power supply will perform within specified limits.

Opto-isolator: Device that provides electrical isolation and a signal path by making an electrical to optical to electrical signal transformation from its input to output terminals. This is accomplished with a light-emitting diode in close proximity to a phototransistor. Opto-isolators are used in the feedback loop to maintain electrical isolation between the input and output of the power supply.

Output impedance: The value of a fictional resistor in series with an ideal voltage source that would give the same magnitude of ac voltage across the supply terminals as observed for a particular magnitude and frequency of alternating current (Chapter 6).

Overcurrent protection: See *Current limiting circuit.*

Overshoot: The amount, measured as a percent of nominal, by which an output exceeds its final value in response to a rapid change in load or input voltage. It is important at turn on and following a step change in load or line voltage (Chapters 4 and 6).

OVP (overvoltage protection): A protection mechanism for the load circuitry that does not allow the output voltage to exceed a preset level. In most cases, the output voltage is reduced to a low value, and the input power must be recycled to restore the power supply output (Chapter 7).

Parallel operation: The ability of power supplies to be connected so that the current from corresponding outputs can be combined into a single load (Chapters 4 and 7).

PARD: Acronym for "periodic and random deviation" and used as the specification term for ripple and noise. Ripple is the unwanted portion of the output harmonically (periodically) related in frequency to the input line and to any internally generated switching frequency. Noise is the unwanted, aperiodic output deviation (Chapters 4 and 6).

Pass element: The active circuit element, typically a transistor, that forms the output power stage of a linear power supply (Chapter 3).

Peak changing: A rise in voltage across a capacitor caused by the charging of the capacitor to the peak rather than rms value of the input voltage. This generally occurs when a capacitor has a high discharge resistance across it and large ripple and noise or spikes on its input line. In a switcher, this parameter affects minimum load (discharge resistance) conditions on each output required to maintain regulation.

Peak transient output current: The maximum peak current that can be delivered to a load during transient loading conditions, such as electric motor starts.

Phase-controlled modulation: A circuit used in switching regulators where the operating frequency is held constant (typically 60-Hz line frequency) and the phase angle at which the control elements are turned on is varied, controlling both line and load changes with minimal dissipation (Chapter 2).

Pin fins: Type of heat sink that uses pins in place of conventional extruded fins (Chapter 5).

Postregulator: Usually a linear regulator used on the output of a switching or ferro power supply to improve overall (load) regulation (Chapter 3).

Power conversion: The processing of medium-quality electric power delivered by utilities to make it compatible with the needs of sensitive electronic circuits (Chapter 2).

Power factor: The ratio of actual power used in a circuit to the apparent power. Power factor is the measure of the fraction of current in phase with the voltage and contributing to average power.

Power-fail detect: A circuit that senses the dc voltage across the input capacitors of a switching power supply. Should the ac input line fail, it senses an abnormally low dc level across the capacitors and provides an isolated TTL output signal warning of imminent loss of output power (Chapters 4, 6, and 7).

Power supply: The common term for electronic devices that provide dc output voltages when powered by an ac primary source.

Preregulator: A regulator circuit (usually a nondissipative type) that provides a line-regulated output, which is then processed by a second regulator, the postregulator, which provides load regulation (Chapter 3).

Programming: The capability of controlling the voltage of each output (Chapter 4).

Pulse-width modulation (PWM): A circuit used in switching regulated power supplies where the switching frequency is held constant and the width of the power pulse is varied, controlling both line and load changes with minimal dissipation (Chapter 2).

Push–pull converter: Used in switching power supplies where the main switching circuit uses two transistors operating in push–pull. The main advantage is simplicity of design (Chapter 3).

Rated pulse power: The maximum power that may be delivered by the power supply on a pulse basis. The rated pulse power usually averages out to the maximum continuous output power.

Recovery time: The time required by a transient overshoot or undershoot in a stabilized output quantity to decay to within specified limits.

Redundancy: The ability to connect power supplies in parallel so that if one fails the other will provide continual power to the load. This mode is used in systems when power supply failure cannot be tolerated (Chapter 7).

Reference: A known stable voltage to which the output voltage is compared for the purpose of stabilizing the output voltage.

Regulator: The part of a power supply that controls the output voltage. In most cases, the regulator acts to stabilize the output voltage at a preset value (Chapter 2).

Remote on–off: See *Inhibit.*

Remote sensing: A method of moving the point of regulation from the output terminals to the load. Compensates for *IR* drops in the power distribution bus (Chapters 4, 6, and 7).

Response time: The amount of time (μs) it takes for an output to react to a dynamic load change and settle within some tolerance band following the load change (Chapters 4 and 6).

Return: An arbitrary name for the common terminal for all the outputs. It carries the return current of all the outputs.

Reverse voltage protection: The ability of a power supply to withstand reverse voltage at the output terminals when hooked up in the reverse polarity (Chapter 4).

RFI (radio-frequency interference): See *EMI.*

Ripple: The periodic ac noise component of the power supply output voltage. See *PARD.*

Second breakdown: Most common failure in the power transistors of switchers; it is caused by the coincidence of high voltage and current levels when the transistors are turned off. Its effects are irreversible, almost instantaneous, and fatal. It can be controlled through proper circuit design (Chapter 6).

Semiregulated output: A secondary output on a multiple-output power supply that receives line regulation only (Chapter 3).

Sequencing: Controlling the time delay and order of output voltage appearance and dropout upon power supply turn on and turn off (Chapter 4).

Series regulator: A linear regulator in which the active control element (transistor) is in series with the load.

Schottky diode: A diode device that exhibits a low forward voltage drop (0.6 V) and fast recovery time. This type of diode is especially useful at high current, low voltage (typically 5 V dc), where low losses and high switching speed are important.

Short-circuit protection: See *Current limiting circuit.*

Shunt regulator: A linear power supply in which the active control element (transistor) is in parallel with the load (Chapter 3).

Slave: The unit in a master–slave paralleling scheme that is controlled by the master unit. See *Master.*

Snubber: A network containing a resistor, capacitor, and diode used in the switching power supplies to trap high-energy transients and protect sensitive components.

Soft start: Input surge-current limiting in a switching power supply where the switching drive is slowly ramped on (Chapter 4).

Stability: The change in output voltage that occurs at constant load, ac input, and temperature after a given period of time following warm-up. This effect is related, in part, to internal temperature and aging effects (Chapters 4 and 6).

Standby current: The input current drawn by any power supply under minimum load conditions.

Static load: A load that remains constant over a given time period. It is usually specified as a percentage of full load.

Stefan–Boltzmann law: A law of thermodynamics that describes the rate of emission of radiant energy from the surface of a body (Chapter 5).

Step change: An abrupt and sustained change in one of the influence or control quantities (e.g., load current).

Stress-aging: The process of subjecting a completed power supply to a variety of stresses to force the occurrence of all burn-in failures (Chapter 6).

Switching frequency: The rate at which the source voltage is switched in a switching regulator or chopped in a dc-to-dc converter.

Switching regulator: A high-efficiency dc-to-dc converter consisting of inductors and capacitors to store energy and switching elements (typically transistors or SCRs), which open and close as necessary to regulate voltage across a load. The switch duty cycle is generally controlled by a feedback loop to stabilize the output voltage (Chapter 2).

Temperature coefficient: The average percent of change in output voltage per degree change in temperature with load and input voltage held constant (Chapters 4 and 6).

Thermal protection: Protection via a thermally actuated switch that interrupts the operation of a power supply if the internal temperature exceeds a predetermined value (Chapters 4 and 6).

Thermal regulation: See *Temperature coefficient.*

Thermistor: A device with relatively high electrical resistance when cold and almost no resistance when at operating temperature. Thermistors are sometimes used to limit inrush current in off-line switchers (Chapter 3).

Transformer: A magnetic device that converts ac voltages to ac voltages at any level. An ideal transformer is a lossless device in which no energy is stored and that requires no magnetic current.

Transient: A temporary and brief change in a given parameter. Typically associated with input voltage or output loading parameters (Chapter 4).

Transient response time: The amount of time taken for an output to settle within some tolerance band, normally following a stated change in load (Chapters 4 and 6).

UL (Underwriters' Laboratories): An independent, not-for-profit organization testing for public safety in the United States. UL recognition is required for equipment used in some applications (Chapter 8).

Undershoot: The amount by which an output falls below its final value in response to a rapid load change (Chapters 4 and 6).

UPS (uninterruptible power supply): A device designed to supply power in the event of temporary or permanent loss of ac line power. Often these supplies will operate with either an ac line input or dc battery backup input.

VDE (Verband Deutscher Elektrotechniker): A West German organization testing for public safety; similar to UL in the United States (Chapter 8).

Warm-up drift: The change in output voltage that occurs during warm-up from turn on of a cold supply until about 30 minutes after turn on. Warm-up drift is measured at constant load, ac line, and ambient temperature and is primarily due to internal components reaching thermal equilibrium (Chapter 6).

Warm-up time: The time needed, after turn on, for the power supply to reach thermal equilibrium with a constant load. Usually estimated to be about 30 minutes.

REFERENCES

Chapters 2 & 3:

Coughlin, Vince. "Where Switching Power Fits In," *Digital Design*, July, 1981, pp 55–58.

Ginsberg, Gerald L. "A Users' Guide to Selecting Power Supplies," *Electronic Packaging And Production*, August, 1980, pp 63–72.

Haver, R. J. "A Designer's Guide To Switching Power Supplies," *Powerconversion International*, July/August, 1982, pp 45–51.

Margolin, Bob. "Modular Power Supplies, Just What Are They?" *Electronic Products*, March 1980, pp 61–65.

_____. "Switchers Pursue Linears Below 100W," *Electronic Products*, September, 1981, pp 45–48.

_____. "Those Sneaky Switchers," *Electronic Products*, March, 1980, pp 39–41.

Snigier, Paul. "Power Supply Selection Criteria," *Digital Design*, August 1981, pp 31–41.

_____. "Special Report: Power Supplies," *Digital Design*, February, 1980, pp 50–62.

Teja, Edward R. "Switching Power Supplies," *EDN*, March 17, 1982, pp 114–128.

_____. "Designer's Guide to Switching Power Supplies," *Digital Design*, January, 1981, pp 66–70.

Chapter 4:

Brown, L. E. "Specifying Switching Power Supplies," *Electronic Products*, November, 1981, pp 51–53.

Landon, Bob. "Myth—Holdup is Free With SMPS," *Powerconversion International*, October, 1918, pp 72–80.

Pfingsten, John A. "Switching Power Supplies," *Electronic Products*, January, 1980, pp 71–72.

Snigier, Paul. "Principles of Designing and Specifying Power Supplies," *Digital Design*, November, 1981, pp 52–57.

Chapter 5:

Drummer, G. W. A. and Griffin, N. B. *Electronics Reliability—Calculation and Design.*

Grossman, Morris. "Pins Cool Better Than Fins And Push Up Power Supply Ratings," *Electronic Design*, July 19, 1978, pp 15–16.

Koetsch, Philip. "MTBF—Rubber Yardstick or Reality?" *Electronic Products*, March, 1980, pp 48–49.

Margolin, Bob. "Military Power Supplies, Are They Really Hirel Products?" *Electronic Products*, August 18, 1982, pp 71–75.

Sass, Forrest. "Keep Cool and Live Longer," *Electronic Products*, March 1980, pp 55–58.

Wilson, Edward A. "System Modeling for Thermal Design of Supplies and Systems," *Digital Design*, April, 1982, pp 32–41.

Chapter 6:

Boschert, Robert J. "Reducing Infant Mortality in Switchers," *Electronic Products*, April, 1981, pp 53–54.

Chamberland, David A. "Automatic Testing of Power Supplies in a Production Environment," *Powerconversion International*, October, 1981, pp 59–63.

Duffy, John A. "State-of-the-art Concepts Applied to Power Supply Testing," *Evaluation Engineering*, April, 1982, pp 22–25.

McLaughlin, D. "Practical Considerations When Testing Switching Power Supplies," *Powerconversion International*, May 1981, pp 34–36.

Parker, Ray. "Economical Power Supply Testing," *Digital Design,* December, 1981, pp 14–18.

Seiler, Rolf. "Quality Assurance of Military Power Supplies is Adaptable to Commercial Applications," *Evaluation Engineering,* September, 1982, pp 32–39.

Swoboda, Jack. "Tips on Testing High Power Switchers," *Electronic Products,* April, 1981, p 41–44.

Teague, Dane. "Automating Power Supply Testing," *Electronic Products,* April, 1981, pp 63–65.

_____. *Test Procedures,* National Power Products, Pompano Beach, FL, pp 4–5.

Chapter 7:

Bailey, J. S. "Distributed Power Sources: Key to Successful Distributed Control," *Control Engineering,* July, 1978, pp 31–34.

Budzilovich, Peter N. "Picking the Proper Power Supply," *Electronics,* June 16, 1981, pp 105–152.

Centofani, E. B., Hansel, A. B., Lioio, P. N., and Natarajan, T. "Optimizing Minicomputer Power Subsystem Design," *Computer Design,* September, 1980, pp 133–140.

Hirschberg, Walter J. "Power Supply Bells and Whistles," *Electronic Products,* April, 1981, pp 47–50.

Koetsch, Philip. "The Perils of Paralleling Power Supplies," *Electronic Products,* April, 1981, pp 37–40.

Kompass, E. J. "Power Supplies Spreading Thinner," *Control Engineering,* July, 1978, p 25.

Lee, Lawrence. "Power Source Distribution," *Electronic Products,* June 30, 1981, pp 43–44.

Chapter 8:

Shepard, Jeffrey D. "Switching Power Supplies: the FCC, VDE, and You," *Electronic Products,* March 1980, pp 51–53.

Sonderby, Iver. "EMI, the FCC, VDE and You," *Electronic Products,* June 1980, pp 45–46.

_____. *International Safety and Emissions Handbook,* ACDC Electronics, Oceanside, CA 1891.

_____. *The FCC and You,* Sierracin/Power Systems, Chatsworth, CA, 1981.

Chapter 9:

_____. *A "Make" or "Buy" Analysis for Power Supplies,* Hewlett-Packard, Palo Alto, CA 1977.

_____. *Power Supply Specifying Guide,* AMP Inc., Harrisburg, PA, 1980.

Chapter 10:

Davis, S. "Top 10 Power Supply Makers Outpaced Industry Growth," *Electronic Business,* August, 1979, pp 62–68.

Grossman, Morris. "Focus on Linear Power Supplies: A Mature Technology Holds On," *Electronic Design,* December 6, 1980, pp 123–130.

_____. "Focus on Switching Power Supplies: Advances Fuel High Growth Rate," *Electronic Design,* March 5, 1981, pp 147–157.

Hachmeister, R. "Power Supply Technology Advances to Meet New Market Needs," *High Technology,* April, 1980, pp 10–12.

Hersom, W. and Shepard, Jeffrey D., "Switching Power Supplies," *Digital Design,* December, 1979, pp 74–78.

Shepard, Jeffrey D. "The Changing Power Supply Scene," *Computer Design,* January, 1981, pp 130–137.

General:

Gottlieb, Irving M. *Regulated Power Supplies,* Howard W. Sams & Co., Indianapolis, IN, 1979.

Hnatek, Eugene R. *Design of Solid State Power Supplies,* Van Nostrand Reinhold Co., New York, NY, 1980.

Pressman, Abraham I. *Switching and Linear Power Supply, Power Converter Design,* Hayden Book Co. Rochelle Park, NJ, 1977.

Tharp, Stephen J. "Evaluating Power Line and Power Supply Performance in Computer Systems," *Digital Design,* February 1980, pp 28–36.

_____. *Application Notes,* Acme Electric Corp., Cuba, NY, 1982.

_____. *Application Notes,* Kepco Inc, Flushing, NY, 1981.

_____. *Application Notes,* Lambda Corp, Melville, NY, 1982.

_____. *Power Supply Application,* Power General, Canton, MA, 1980.

_____. *Power Supply Engineering Information,* Sorensen Company, Manchester, NH, 1982.

_____. *Switcher Book, A Switching Regulated Power Supply Handbook,* LH Research Inc., Tustin, CA, 1981.

_____. *Technical Data,* ACDC Electronics, Oceanside, CA, 1980.

_____. *Technical Data,* Power/Mate Corp, Hackensack, NJ, 1982.

Wood, Peter. *Switching Power Converters,* Van Nostrand Reinhold Co., New York, NY, 1981.

INDEX

P

R

S